D0079029

the grow home

the grow home

Avi Friedman

McGill-Queen's University Press
Montreal & Kingston • London • Ithaca

ISBN 0-7735-2168-2 (cloth)

Legal deposit second quarter 2001
Bibliothèque nationale du Québec

Printed in Canada on acid-free paper

McGill-Queen's University Press acknowledges the financial support of the Government of Canada through the Book Publishing Industry Development Program (BPIDP) for its activities. It also acknowledges the support of the Canada Council for the Arts for its publishing program.

Canadian Cataloging in Publication Data

Friedman, Avi
The Grow Home
Includes bibliographical references and index.
ISBN 0-7735-2168-2 (bnd)
1. Dwellings – Design and construction. 2. Dwellings – Design and construction – Economic aspects. Bibliography

TH4812.F74 2001 C00-9009949

This book was typeset by David LeBlanc
in 9.5/13.5 Meta

CONTENTS

PREFACE

In June 1990 the Grow Home was inaugurated on the campus of McGill University. The narrow-front unit was constructed across from the grey limestone School of Architecture building where it was designed. A year earlier Witold Rybczynski and I had co-founded a new program at the school with a mandate to teach, research, and advocate the idea that design solutions to the affordable housing crisis of the time were genuinely feasible. We hoped that our Grow Home – a single slice of a row of other such homes – would demonstrate that trade-offs made by home buyers, such as purchasing a smaller unit, would get them through the door of their first home. We also hoped that private-sector builders would be interested enough to respond and even profit from this market niche.

The freestanding structure was an outcome of a year-long design process linked to a questionnaire that was prepared at the request of the *Montreal Gazette* and appeared in the newspaper in February 1989. The questionnaire was informal, and the fact that only sixty people bothered to respond made it hardly representative or statistically valid. Even so, it provided insight into the opinions of people who did not own a home but would have liked to.

What sort of home would respond to the stated preferences of these rental tenants? First of all, it would have to be reasonably close to downtown. This would not only reduce the cost of commuting for an individual family but would also slow urban sprawl. In the Montreal tradition, a home close to downtown suggests a rowhouse, a housing form that reduces the cost of land and infrastructure. Next, for a family with a combined annual income of $50,000 (the category to which the majority of the questionnaire respondents belonged), a home with a price under $70,000 would be affordable. Such a price was far lower than that of the available single-family homes. Finally, the question that received the largest positive response was, “Are you ready to do some of the finishing work yourself?” This suggested an approach that we designated as a “grow home” – a partially finished house where certain components could be finished by the owners and where some spaces (either on the lower or upper levels) were left unpartitioned for future completion.

Those questionnaire responses helped us to form an image of future occupants and establish basic design principles. A grant from Dow Canada moved the design from the drawing board to the Modulex factory floor where the unit was constructed in panels, and ultimately to the university's grounds where the home was assembled. On 20 June 1990 the Grow Home was ready for inspection.

We were naturally anxious as we awaited visitors' impressions of the design. They walked through the 1,000-square-foot, two-storey unit, tastefully yet frugally furnished, and observed the dining area at the front of the lower level which shared space with simple, Shaker-style, maple kitchen cabinets. They carried on down the hallway fitted with white, twelve-inch-deep cupboards and a shelving system, paused at the simple bathroom tiled with black-and-white checkered linoleum, and came to the living room at the rear with its double French doors through which they could see a wooden deck and pergola. Climbing the stairs to the upper level, at one end they saw a baby's room with a crib, large nursing chair, and toys scattered on the floor, and at the other end an unpartitioned space with a double bed, desk, and large pine clothes cupboard on greyish-blue carpet extending all the way to double doors that opened onto a balcony at the front of the house. They used a set of stairs built along the side wall to leave the Grow Home, wondering as they looked around once more whether this slim grey house was the solution to the affordable housing crisis about which they had heard so much. We waited nervously, not knowing the effect of our design.

The front page of the *New York Times* Homes section ("Can Small Be Cheap and Also Beautiful?") and ABC News's satellite dish broadcasting live from the McGill campus on *Good Morning America* alerted us that we were on to something big. Perhaps we had even underestimated the extent of the housing crisis and how urgently solutions were needed, not to mention the eagerness of the media to write about it all. The month-long demonstration of the Grow Home brought an avalanche of interest: articles, interviews, and telephone inquiries from near and far. It also generated opinions and views about the merits of a narrow-front, Grow Home strategy – a debate which we had hoped the demo unit would create.

The uneventful six months that followed the dismantling of the unit were a letdown after the euphoric days of June. Our best efforts to find a real developer with a real site were unsuccessful. Until, that is, Leo Marcotte walked into the school. An owner of a small building firm in east-end Montreal, Marcotte came to participate in a design studio project. He had never heard of the Grow Home, and when we told him about it, he was sceptical of our claims that the unit could be constructed for $40,000 or for approximately $75,000 including land. But he was willing to re-examine our estimates. The fact was he did not have much else to do. Montreal was in the grip of a deep recession, and housing starts had plummeted. When he called later to tell us that he found our numbers to be accurate and that he had decided to take the plunge and place an advertisement in the weekend edition of a local paper, we were mightily heartened. None of us had any idea what the outcome would be.

Marcotte took a risk and won. Twenty-four units were sold in a weekend. The entire development of eighty-seven homes to be built at Pointe-aux Trembles was sold in two weeks. The buyers – most of them renters – could not pass up the opportunity to buy a two-storey unit with a basement for a monthly mortgage payment that would be lower than what they had been paying in rent. They purchased their first homes without even seeing a model unit, completely unheard-of in the Montreal housing market.

In the depressed market of the time, the news of Marcotte's success travelled fast. Other builders came to the site to observe, learn, and copy. Developments started to spring up in and around Montreal. By the end of the first year, about one thousand units were built. The concept has not lost its appeal since; to date over ten thousand units have been built in Canada. The Grow Home has inspired developments and projects across North America and earned numerous accolades and awards. It was more than a mere academic experiment and a local success in affordable housing. The Grow Home demonstrated that efficiently designed, well-built homes can be affordable and also appeal to buyers' aspirations and, as such, should be considered as a valuable option among other major solutions. I had also recognized by then that a large laboratory had been constructed: communities, homes, and people worth studying. What followed was a decade of evaluations, research, and design that led to more ideas, developments, and articles upon which this book is based.

The book begins with a sketch of a society at the twilight of the twentieth century and the dawn of the twenty-first. I have attempted to explain new demographic compositions, economic realities, lifestyles, and the effect that they all have on home life and the search for a new housing paradigm. An outline of our design principles then ties the Grow Home ideas to earlier architectural examples and demonstrates how those ideas have been modified to fit our own times. In the campus prototype, a variation of these concepts illustrated in the following chapter, the reader will discover which factors influenced the design of spaces, rooms, and details. Response to the demo unit is presented as the results of a questionnaire that visitors were asked to fill out at the end of their house tour in June 1990. The transformation of the prototype into a widely built unit is described next, giving the builders' side of the story, based on extensive interviews with seven of them and on the data they provided. This chapter provides a close-up of the modus operandi of a small homebuilding firm and its quest for a bestselling model. Then, 196 Grow Home occupants provide insight into the minds of first-time home purchasers: their aspirations, the appeal of specific features for them, and the reasons behind their decisions to buy. The following chapter describes their dwelling experience; visits to 140 Grow Homes furnished the opportunity to document and ask about work owners did in spaces that were originally left unfinished – what they chose to build and how they did it.

The next chapters mark a slight shift of approach. From a chronological description of events, I move on to focus on issues unique to the Grow Home that further explain the magnitude and potential of the design. The many built units inspired a factory-built home: we attempted to design a prefabricated version of the Grow Home in which numerous interior variations could be produced using a limited number of interior and exterior panels. The concept of the Grow Home was extended to the developing world in La Casa a la Carta, which I designed in 1997 with Mexico (where it was built) in mind. In countries other than our own, a small structure is often merely the foundation upon which additions are built according to available means and needs.

During the design phase of the Grow Home we realized that building small not only reduced construction cost but also conferred an environmental advantage. Resource conservation and significant energy savings were obtained through modular design, floor stacking and the joining of units. These strategies demonstrate the green aspect of the Grow Home and they are outlined in this chapter.

The Grow Home was designed to be part of a row and to be constructed in a community of rowhouses. In the years following the construction of the campus model, I had the opportunity to design several such communities. This posed a particular challenge, one directly related to higher density. I found that attention must be paid to (among other topics) roads and parking, open spaces, and the avoidance of facade repetitiveness. The chapter titled "Neighbourhoods with a Sense of Scale" illustrates some of these design patterns. The book's wrap-up chapter reflects on the entire experience and – more significantly – looks forward to the future of the Grow Home concept and its ability to continue as a viable housing solution by describing a subsequent design – the Next Home.

In 1999 I received on behalf of my colleagues the United Nations World Habitat Award for the achievement of the Grow Home. The jury cited the potential of the idea for replicability elsewhere as the main reason for the award. I believe that the Grow Home design has only just begun its life and that many more chapters remain to be added in the future.

ACKNOWLEDGMENTS

This book is based on design and research spanning approximately ten years. The conception and design of the Grow Home were a collaborative effort with Witold Rybczynski. The development of the demonstration unit also involved the architect Susan Ross. In the following years, I worked closely with Vince Cammalleri with whom I co-authored several papers on which some of the chapters in this book are based. There were other participants in the research and in the organizations which funded it. I would like to acknowledge and thank them here:

Christine Von Niessen for her assistance with the research on postwar housing innovation.

Aurea A. Rios for her assistance with the research on the post-occupancy modification of Grow Homes by their owners, which formed part of the research for her Masters of Architecture degree.

Dow Canada Inc. for the financial support of the research, development, and construction of the Grow Home demonstration unit. Their support was the beginning of it all.

Modulex Inc. and Ikea Canada for constructing and furnishing the Grow Home demonstration unit. Josée Lamonte and Donald Chan for their assistance with processing the visitor questionnaire.

The many builders and, in particular, Leo Marcotte, for implementing the Grow Home concept and for sharing their experiences with us.

Canada Mortgage and Housing Corporation and especially Donald Johnston for enabling us to conduct the post-occupancy studies. Virginie Desjardins for processing the questionnaires.

Jacques Trudel and André Poitras from the Société d'habitation du Québec and Peter Russell from Canada Mortgage and Housing Corporation for their interest and valuable feedback during the prefabrication of the Grow Home. Paola DeGhenghi for her work in the production of CAD drawings and development of standardized design alternatives and Ann Drummie for her assistance in the survey and evaluation of panel systems.

Canada Mortgage and Housing Corporation and Société d'habitation du Québec for their support of the La Casa a la Carta research and construction. Also Chen Lin, Mylène Poirier, Nadia Meratla, Miljana Horvat, Miguel Rojano, and Susan Parmley for their contributions.

François Dufaux, David Morin, and Rosanne Howes for their diligent work on the design of the communities presented here.

Michelle Kwok for her many meticulous drawings in this book and Henry Tsang for compiling all the graphic material.

Philip Cercone, Aurèle Parisien, Joan McGilvray, and Susanne McAdam of McGill Queen's University Press for their spirited encouragement and help.

This book could not have been accomplished without the assistance of David Krawitz. Over the past six years David has edited many of the papers on which chapters in this book are based as well as the chapters themselves. His dedication and attention to proper representation and accuracy are truly admirable.

the grow home

1 DIFFERENT TIMES, DIFFERENT HOMES

When the Grow Home was introduced in June 1990, it stood in marked contrast to the homes typically built by Montreal developers. "Why is it small?" people asked. "Why is it narrow? What does the "Grow" mean?" Still other visitors wondered whether the Grow Home was meant to be a trend-setter in architecture.

The design was an architectural response to many changes that had influenced society in the preceding decades, changes that had a profound effect on home life. One indication of the scope of these changes is reflected in the public perception of the traditional family in the years immediately following the Second World War in comparison with the current conception of the family. The postwar image of the breadwinner father, homemaker mother, and a minimum of three dependent children living together in a detached house was so pervasive that homebuilders could easily – and successfully – view the bulk of their potential clientele as a homogenous market. In the eyes of builders as well as for many North American citizens, single people and single parents rarely bought houses. The majority of people got married in their early twenties and immediately began their search for a house. General expectations of hearth and home were universally shared. Simple housing formulas worked, which subsequently justified maintaining the traditional manner of designing homes for a dominant majority with common residential habits.

The telecommunications revolution has brought the furthest corners of the world right into our homes, to television watchers and computer users alike. The widespread practice of reliable birth control has enabled people to regulate and limit the size of their families more than ever before. Economic recessions and restructuring have taught both earners and the unemployed the lessons of prudence and financial impermanence. The upheaval in social values ushered in during the 1960s has led to a new era of relaxed morality and lifestyle accompanied by corresponding reactions of outrage and conservatism. The result has been radical shifts in the way people live and form households, work and enjoy their leisure, grow old and die. The homes themselves were frequently central in these changes, which affected most aspects of domestic life: interactions between family members, consumption patterns, the type of room arrangements people chose, and the actual use of these rooms. All homes cannot be examined as though they are the same – there are city homes, suburban homes, rural homes. Wealthy people live in some, middle-income and low earners in others. Yet the postwar changes that so affected home life influenced the vast majority of people in most homes.

The Grow Home was in part an architectural interpretation and a possible response to these recent changes. The quest for a new architectural paradigm was not uppermost in our minds. It was rather a process of reflection on current phenomena, an examination of coming trends, an assessment of case histories, and a composition of construction strategies: strategies that would make homes affordable to those people who were unable to purchase them as a result of those same societal changes we had been noticing. Cost, however, was not the only consideration. We focused on the design of a home that would fit the everyday needs of its occupants when they moved in and that would let them modify the home as their needs and means evolved.

So what were these changes and how did they influence our design? Let us begin with society itself. The decline of the traditional nuclear family and the emergence of other household formations are at the heart of the social shift. For some, obtaining higher education became an entry passport to the job market, and this delayed the age of marriage and the start of new families. Newlyweds began their search for a new home later. When they finally did enter the market, they soon discovered that the cost of a monthly mortgage payment – combined with the repayment of their student loans – required a change of plans, primarily for women: work as late as the biological clock allows, have fewer kids, and return to work as quickly as possible after the birth of a child. Between 1975 and 1999 the birth rate per 1,000 people dropped from 15.5 to 11.2 (Statistics Canada 2000a). These same first-time home buyers also had to settle for a less expensive home. This meant a smaller unit, fewer amenities, a partially finished space, or a move further from the city centre to buy a home on inexpensive land. The small space was a compromise for new householders since they knew that there would be fewer people in the family, and besides, there would always be a second and third home down the road. As they stood in a developer's model home, making up their minds about whether or not to buy, whispering to one another, they knew that they would have to tone down their expectations and dreams. They would have to change a few habits and alter their lifestyles.

One of the significant causes of change in the lifestyle of North Americans is the continuing increase in the proportion of women in the labour force. The ramifications of greater numbers of working women are numerous: women in the labour force must schedule their time to balance paid work, domestic work, and leisure; single mothers have the added burden of independently providing for the care of their children in their absence; married men have to share more of the family and household tasks. And all must ensure that after the responsibilities and chores of their workday lives are done, they have sufficient time for their families, friends, and themselves – a recent notion born of the necessity of coping with a seemingly endless list of daily activities.

As a result of the high cost of raising children and the need for a quick return to the labour force after childbearing, households with young children often find they have a reduced need for space in the home. They also have less use for low-priority space (for example, infrequently used rooms such as the dining room and outdoor areas) and a greater requirement for low-maintenance housing (Miron 1981). Many homes are often empty during the day, while morning "rush hours" put pressure on bathrooms, closets, and the kitchen. Efficiency is a fundamental element in the configuration of the dwelling unit. The necessity of accomplishing a great deal before leaving the home in the morning, for example, influences the purchase of appliances and their location. Traditional tasks that are commonplace to any home, such as cooking and laundry, have evolved to take on new features and manifestations. The central location of the washer and dryer in the home, frequently in or near the kitchen, is often chosen not only because a basement may not be an available option but because the concentration of household chores within a single area increases efficiency for the household members performing the tasks.

The number of non-traditional households has also expanded with an increase in common-law households. Statisticians, demographers, sociologists, and other researchers share the agreement that there no longer exists any universal definition of "family." Social, economic, and technological changes and shifting values in recent years have influenced the structure of families so that new forms have emerged. Housing designers and providers must reformulate their conception of what constitutes a household to comply with arrangements that include married and unmarried unions (male and female and same sex) in order to design, construct, and market homes that appeal to the wide spectrum of housing user types.

A household type which has gained in numbers over the years is that composed of one person only. Contrary to previous conceptions, nowadays not only do many singles purchase homes on their own before marriage but many buy homes without the specific intention of marrying in the conceivable future. Many of these young single home buyers, incidentally, are not male: more women than ever before own homes. Builders frequently neglect to orient their products towards single people and are thus excluding a considerable portion of first-time buyers.

The decision-making process of single people in buying homes follows a particular logic. For many, entry into the housing market is stimulated by the realization that they may remain single for some time and that they are not building equity while they are paying rent as tenants. Sharing a floor with others in an apartment building with a long elevator ride and dimly lit corridors begins to lose its appeal. Once the idea of owning a home is raised, it is hard to dispel the image of reading the weekend paper on a lounger on the backyard deck. But suburban developments with McMansions are not designed for single earners, especially those with modest incomes. Besides, a single person is not inclined to live in a house with huge front and back lawns, half a dozen rooms, marbled bathrooms, a kitchen the size of a football field, and a three-car garage. And who would pay the cleaning staff? An affordable, small, easy-to-maintain home never used to be available in these suburban developments, nor anywhere else for that matter. A home like this represented an untapped market niche.

Single-parent families were in an even tougher predicament. Accounting for one-tenth of all North American households, this family type – headed primarily by women – faced considerable financial challenges and space limitations. The need to work, raise children, and put their own lives together were awesome enough tasks in themselves. Just like single people, most families headed by a single breadwinner could not afford a large home. They also had the additional challenge of allocating their available space. In families with shared custody, the kids were likely to spend every second weekend with the second parent. Two small homes now had to be created with children's bedrooms in both. In homes where the mom or dad had begun to date again, there was a further complication in the assignment of space, this time over a more sensitive issue. As well, the proximity of a parent's bedroom to adolescent kids might not be acoustically ideal, especially when one of the kids had a friend to sleep over. Designing stacked affordable units with bedrooms on different levels or apart from one another was an appropriate solution for such households. Ahrentzen (1991) and Leavitt (1991) studied housing arrangements of non-traditional families. They argued that a dwelling unit designed for pre-occupancy choice of interior layout and situated within a development with mixed types of housing would allow for a row of structures that could be attractive and suitable to many living arrangements and family compositions. By incorporating flexibility directly into the housing strategy, each unit could be adapted to the individual requirements of either nuclear or non-traditional families so that both could easily be accommodated within the same development without any social or physical demarcation. The need to balance the options available to all households is of paramount importance. Builders cannot anticipate the precise nature of all the buyers who visit their sites, and they will better serve themselves as well as potential purchasers by incorporating an open-ended design strategy that facilitates and expedites their decision-making abilities. Thus, in addition to affordable price, an essential consideration was adaptability to the needs of a wide range of household types. One way to proceed was to design and build a narrow unit and eliminate the need for structural bearing partitions that would inhibit flexibility.

Our ideas required not only a flexible design but a flexible mind-set on the part of financial institutions towards first-time borrowers. Unfortunately, this flexibility was not to be found. Economic changes – a result of fundamental global shifts – swept through lending institutions and changed many practices, which had a direct influence on home purchases. Young people were the most surprised of all. It was if a carpet had been pulled from under their feet. Upon graduation, they had begun their search for their first real jobs. Attempting to replicate their parents' experience, they assumed they would find long-term, secure employment. They expected to grow with a company and receive a pension at the end of long, dedicated service. When they finally arrived at their first job interviews, however, they listened to human-resource managers using terms like "accountability to shareholders" and "market fluctuations" and "forthcoming mergers." At the end, if they were lucky, they might have been offered a one-year contract job, hardly part of their cherished life plan. Hard work and steady company performance, they were told, could earn them a two-year job extension. It was a bitter pill to swallow, but no different from any of their other interviews. Plans would have to be readjusted. What about getting married? Buying a home?

The meeting with the bank manager did not make decisions any easier. "Employment status?" the manager would ask soon after a formal introduction. "Contract employees." Tapping the borrower approval form on his desk, the manager might repeat that answer, slowly, before announcing, "No good for us. What would happen to the loan if one or both of you were out of work? I'm sorry, but I don't think your loan can be approved."

First-time home buyers found themselves in a bind. The residential mortgage system had been established in times of continued economic prosperity when market trends and foreign competition were not threats and lenders were eager to lend. The system was based on stable employment and the secure ability of borrowers to pay off debt. Our future homeowners had no choice but to modify their expectations. The banker was right, they realized. What would actually happen if one of them stopped working? How could they continue to make hefty monthly mortgage payments on a large home? The options became clearer. Continued renting would not give them a head start in home-ownership. But they could not commit themselves to a huge mortgage, and they would have to buy a smaller home. Their monthly mortgage payment would, they hoped, be equivalent to their present monthly rental payment. If one of them stopped working, they would still have some savings to cover payments until the next job came along.

An undeniable component in the design of an affordable home has become the amount that potential buyers are willing and able to pay month after month. A multitude of design strategies are all aimed at obtaining that magic number. Surprisingly, some of these strategies – such as smaller homes on small lots and with unfinished parts – go against the grain of conventional financial wisdom. Bankers want home buyers to commit themselves to a large loan, as long as it is paid back. In fact, potential buyers are encouraged to borrow. When the modern mortgage system was set up in the years after the Second World War, regulators allowed 25 per cent of a combined household income to be allocated for shelter expenses (mortgage payments, municipal taxes, and heating). It was assumed that the remaining 75 per cent would be spent on other household expenses such as education, old-age savings, and food. The 1950s saw a large number of first-time home buyers lining up outside model homes across North America. As the husband was the only breadwinner in the family, that 25 per cent for shelter would come from his income. Fast forward a decade or two. The 1970s saw the return of the housewife to the labour force, and most families now had two pay-cheques. Regulators did not miss this phenomenon. The allowable shelter allocation was raised to 32 per cent of the new, enlarged, total household income. In real terms, home buyers were allowed to spend much more on housing.

Builders did not miss a beat either. They knew that greater amounts of disposable income meant an improved ability to spend more on a home. There is no comparison between new homes today and those that were constructed at the beginning of the mass-consumption housing market of the 1950s. Back then, the ability of the middle class to own a home was considered one of the greatest postwar achievements. Tract homes with few amenities were meant to satisfy the basic needs of a young family. Such a modest beginning, it was argued, would be a stepping stone towards the acquisition of larger, better-equipped homes. In 1943 an average Canadian home measured $80m^2$ ($800 ft^2$). By 1955 that size had increased to $110m^2$ ($1,100 ft^2$). By the mid-1980s, when the baby boomers came to the market, single-family home sizes were close to $200m^2$ ($2,000 ft^2$). This expansion came hand in hand with a decrease in the size of the average household from 3.9 members in 1961 to 2.5 in 1998 (Statistics Canada 2000b). We have, quite simply, increased the amount of space per person.

Builders did not stop at larger houses. They loaded them with costly objects, features, and gadgets. Contemporary home design reflects this trend. If the postwar house represented the provision of shelter, today's home is a gateway to social status. Within the context of identical income level and demographic make-up, new developments vie with one another to appear different. Paradoxically, they all have the same basic interior and attention-grabbing features. Detailed craftsmanship has been replaced with a huge foyer, high ceilings, tall windows, curving staircases, and fireplaces destined not to be used. A new home is a parade of wonders. Every room has a new feature or gadget to simplify life. Consider, for instance, the kitchen. The microwave oven has led the transformation of the kitchen from a cooking to a processing place. It has also changed the food we buy from raw ingredients to ready-to-serve products, altering our need for storage space. Bigger fridges and freezers – sometimes even a second fridge – have replaced pantries stocked with staples, pickles, and preserves. The size and number of kitchen cabinets have increased to store juice makers, pasta machines, automatic bread makers, popcorn poppers, cappuccino makers, food processors – all the appliances we received as Christmas, wedding, or housewarming gifts or which we bought after watching a cooking show, only to use the appliance once and then bury it at the back of a lower cabinet.

Clever marketing and consumer demand have also transformed the bathroom. This once-humble space now boasts a veritable spa of appliances: whirlpool bath, multi-jet shower, silent-flush toilet, and double-sink counter with a row of theatre make-up lights contend for our attention in an Italian-tiled paradise. The main bathroom is just one of many in such a home, in addition to a couple of powder rooms, one on each level, and an ensuite bathroom in the master bedroom. Builders invented new functions when they had too much space on their hands and too few uses. "Media rooms" were designated to house our electronics. The leftover space next to the kitchen became the "great room" (not be confused with the family room), and the "bonus room" came to life when builders ran out of names. All of these rooms were created in addition to new space put aside for the salient feature of the new economic landscape: the reincarnation of the work-at-home phenomenon.

More and more people are transforming part of their homes into working environments and themselves into home-office telecommuters. The availability, affordability, and accessibility of personal computers, fax machines, modems, and photocopiers have, for many telecommuters, outweighed the expense of rented office space: by working from home, a person can economize on office costs which can then be channelled directly into business-related resources. Other factors causing an increase in the number of home workers are the proliferation of information-based and service-related jobs, more contract work and part-time employment, improvements in computers and telecommunications, the effects of corporate restructuring, the effort to integrate work and home as a means to improving living environments, and the general desire of workers to balance the spheres of family life and work.

Many people, when they decide to modify spaces in the home for work-related use, express the desire to convert an unused room such as the living room Into an office. Such an admission is revealing: the fact that some people consider the living room "unused" space suitable for conversion implies that it may not be unreasonable to offer prospective homeowners the choice of allocation of their future spaces, specifically, the choice of whether they want to replace the living room with a home office or combine the two spaces. The greatest number of future buyers would be served, however, by designing the living space with the capacity to be easily converted, either in whole or part, to an office area and back again, at the discretion of the owner (Gurstein 1995; Senbel 1995). Perhaps the greatest effect of the work-at-home phenomenon is the flexibility that future homes will have to offer to their first and subsequent occupants. Unpartitioned space will initially be a cost-reduction design strategy but may subsequently become a source of revenue as a home office or a space to be used for other family needs as they appear.

Adjusting spaces to a family's changing needs has become easier in recent years because of the unparalleled expansion in home improvement products and their retailing. Huge renovation supermarkets are places where the customer is not only greeted cheerfully at the door but can buy anything from a single nail to an entire home. Products have been simplified to be used by a lay person. No more waiting for days or outrageous charges by a plumber or electrician: do it yourself! The do-it-yourself movement is marked by the proliferation of television shows, videos, books, and magazines targeted at home renovators in order to facilitate participation in this new segment of the housing industry.

Attitudes to home renovation have also changed. In the decades following the Second World War, it was considered appropriate to move to a new and larger house to improve residential conditions and family status. Today it is equally acceptable to improve the current house rather than move. Aside from the financial incentive (avoiding the risk of burdensome costs of a more expensive home), renovation satisfies a multitude of personal impulses for modifying living space, such as the need for change, the ability to express individuality, and the desire to exercise control over a portion – however small – of one's living conditions. Alterations may be made in response to life cycle transitions, such as creating or rearranging space for newborn and young children, converting areas for use by adolescents and young adults, recovering space vacated by departed children, and redefining space when a household member returns to work or study. Other influential factors include changing styles and tastes, narrowing the gap between the current and the "ideal" home, and personal self-expression.

These trends support the concept of design principles whereby buyers are directly incorporated into the process with the option of determining the design best suited to their individual needs. Potential buyers hoping to economize on all aspects of home-related costs may well choose to purchase a unit with unfinished spaces or features, knowing in advance that part of the home-buying process includes the opportunity to accommodate post-occupancy modifications made by the residents at their own convenience.

Buyers of large homes have much to renovate: a big kitchen, several bathrooms, a media room, and many studies are part of the new residential landscape. Not only have the homes themselves expanded but the lots on which they were built have swollen in size. The roads in front of these houses have grown wider, and the utilities installed beneath them to feed and drain the houses have grown in number. The urban landscape and its economics have changed.

The first postwar suburbs were built on inexpensive land. Tract builders purchased large farms and developed communities with streets frequently measuring 12.2m (40 feet) in width. Land costs were secondary to building costs in the overall price of a house. The municipalities in which these developments were built were entrusted with providing the services. Public money was spent on roads, sidewalks, electricity, water, drainage, and fire hydrants. Governments made loans to municipalities, which in turn collected taxes from home buyers. But in the late 1970s cash-strapped governments across the continent stopped lending to municipalities. The construction of services became the responsibility of developers who were also asked to finance it. Reasonably enough, they tacked these costs onto the price of a new home. The cost of infrastructure doubled overnight. Home buyers were now asked to pay up front and immediately for a housing component that, in the past, they paid for over a period of time.

Development standards have not declined. Despite the higher cost, municipalities have enshrined wider roads and large lots in their bylaws and planning codes. It became apparent that if one wanted to reduce costs significantly, reducing those standards would be one possibility. The cost of services in a city like Montreal amounts to $500 per linear foot (.3m) on each side of the street. The owner of a home built on a 15.2m (50-foot) lot (a typical size) would therefore pay $25,000 simply to service that home. Narrowing the lot would lower the service cost substantially, but no municipality was eager to do this. In towns that draw little tax money from industries, residential developments have become cash cows that pay for policing, swimming pools, parks, and bureaucracies. Smaller homes on smaller lots had little appeal for such communities: they simply could not generate enough taxes. Worse, they appeared to threaten the value of the large, expensive homes. The twin phenomena of Not In My Back Yard (NIMBY) and Not In My Term In Office (NIMTIO) reared their heads. Residents who were paying costly mortgage payments on their homes objected to anything that might lower their value – especially small affordable homes. Potential homeowners with lower means became victims not only of a cruel economy but of their fellow citizens. The task of designing affordable homes was complicated by the challenge of having them built in communities that did not welcome them.

Have governments intervened? Not at all. In fact, over the past two decades governments have got out of the business of housing lower-income people. In many countries government-initiated social housing projects were the first to go when harsh budget cuts were introduced. Bureaucrats recognized that there was no wisdom in the government acting as a developer. The construction, sale, rental, and management of housing projects, they reasoned, would be better relegated to the private sector. This was not an ideal choice, since private-sector builders would be involved, and their expertise was in constructing and selling large homes. As designers, we had to develop concepts and designs that would not only be affordable but would also coincide with industry practices and would employ the same products and techniques. We had to bear in mind the future builders of our designs.

Societal changes and their effect on homes and home life focused our attention on necessary design strategies. The next step was to construct the design ideas themselves – that is, to see what should be included in a list of principles that would suit the everyday needs and income levels of the Grow Home's target purchasers. Study of cost-reduction strategies brought to our attention previous architectural attempts to design affordable housing. In the euphoric boom years after the Second World War, returning veterans lined up to purchase homes while builders worked hard to produce them quickly, efficiently, and cheaply. To help the builders meet their targets, innovative production methods, designs, and products had to be introduced. So that buyers could qualify for a loan, a price tag, set by government, had to be established. The design for a target budget and for a particular client or a small range of clients served to inspire our work. Yet one cannot simply adopt and implement old ideas; they must be thoroughly examined and put in a new light and combined with more recent trends and directions. What follows is our thought process, which drew its inspiration from case histories.

The easiest way to reduce construction cost is to build smaller houses. The term "smaller" is open to a wide range of interpretations. Many postwar homes measured between 40 and 80m² (400 and 800 ft²) and frequently accommodated families of six. It is fair to assume today that a home that measures between 100 and 150m² (1,000 and 1,500 ft²) will be considered small by the standard of North American merchant builders, and in some markets even a home of 200m² (2,000 ft²) is viewed as small. Statistics indicate that homes of this size accommodate households composed of an average of three people. The number of people who share a space (also known as the measure of crowding) is therefore what counts most. When builders consider the construction of a new home, they tend to set a target retail price.

Price per square foot leads them to an appropriate unit size for their would-be buyers. A smaller home does not need to mean a reduction in comfort, since all the amenities of a modern home can be included, particularly the wide range of electrical appliances and electronic devices that increasingly define contemporary domestic life. A large portion of the savings in area can be achieved by eliminating waste space and by rethinking conventional standards. Construction costs can be kept under control by balancing housing quality, livable area, and the complexity of construction and finishes. A smaller house has other advantages. It is less expensive to heat as well as easier to maintain. In two-adult households where both adults work outside the home, a more compact home means less housework. Needless to say, this same benefit applies to single-parent families.

Some postwar builders broke ranks with traditional ways of building on both conceptual and technical levels. Consumer demand for housing and expectations ran high: people wanted the latest in technology, planning, building, and labour-saving devices, within established designs (Hudnut 1942). Consequently, the housing industry was forced to deal with two underlying realities: in order to keep costs down, building could no longer be undertaken using the designs and methods employed at the time; and savings in small-house construction would come not from major elements but from meticulous attention to innumerable details (*Architectural Forum* 1948). Because of the limitations in both price and size, architects in the field were forced to redirect their practices away from ornate, decorative, and stylish dream houses. Postwar housing required functional, practical, and economical solutions appropriate to family homes.

Economics, material shortages, technological advances, and changes in consumer expectations were all reflected in the design alterations to the typical, low-cost home in the period from 1945 to 1959. The typical family home before the war was a two-storey structure with a pitched roof and a basement. One example of an already compact traditional style that was scaled down and redesigned for use as a prototype for postwar development was the Cape Cod Cottage that had become the popular symbol of home in the 1930s and '40s. Built in 1949 and designed by Samuel Glaser, the revitalized Cape Cod Cottage (figure 1) could be purchased for under $10,000. Its modernization introduced higher ceilings and added foundations to produce a small separate kitchen, two bedrooms, a living room, and a spacious attic which could be converted to bedrooms at a later date (*Architectural Forum* 1949a).

During the postwar period the application of several strategies kept housing prices low without compromising livability. Designs were kept simple and small, with a square floor plan as the frequent focus of the design strategy. The plan provided a maximum amount of floor space within a minimal amount of wall construction. The introduction of multi-purpose rooms (which served the same function as hallway, dining room, living room, recreation room, and den) also reduced costs. This type of strategy called for a reorganization of traditional house planning. An expanded living room still served its traditional function but took on new roles such as study, dining room, parlour, and play room. The kitchen, now considered

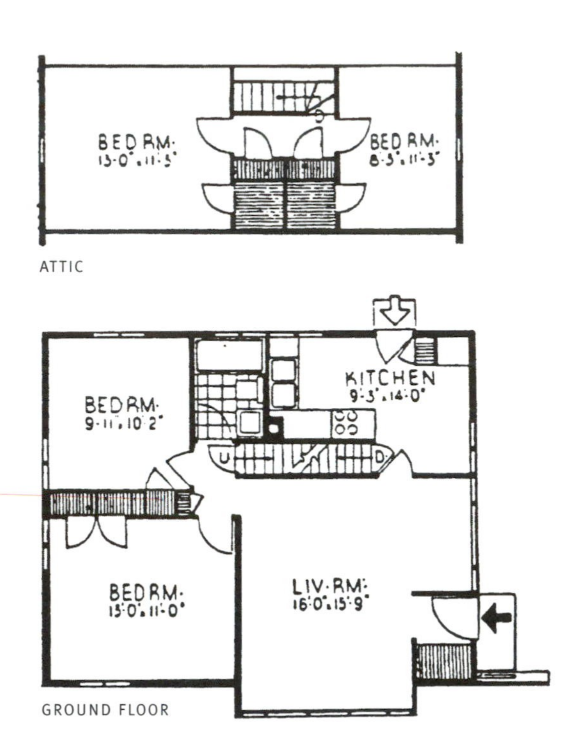

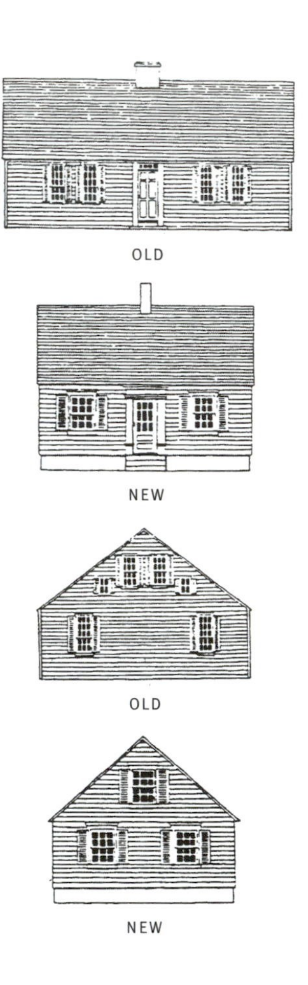

Figure 1 Revival of the popular seventeenth-century design in a more compact form: revitalized Cape Cod Cottage – Samuel Glaser, architect (*Architectural Forum* 1949a)

in the social as well as in the domestic realm of the house, was no longer relegated to the back of the house but was integrated into the multi-functional living area. Converted into a U-shaped work space equipped with appliances and gadgets, a low counter was all that separated it from the living area. Thus the kitchen became a functional, flexible, and efficient space that could accommodate adult entertainment activities as well as supervision of children's play. One example of such an innovative house was built in 1950 in Seattle, Washington, designed by Paul Hayden Kirk (figure 2). The use of a central mechanical core that comprised the kitchen, bath, and storage closets into the small plan (82.5m² / 825 ft²) eliminated the need for fixed partitions or doors. The total cost was $10,500 and was achieved using standardized post and beam lengths (Ford and Creighton 1954).

The scarcity of interior space and the changing spatial needs of the family served to reduce the emphasis on fixed features such as walls. Rooms could

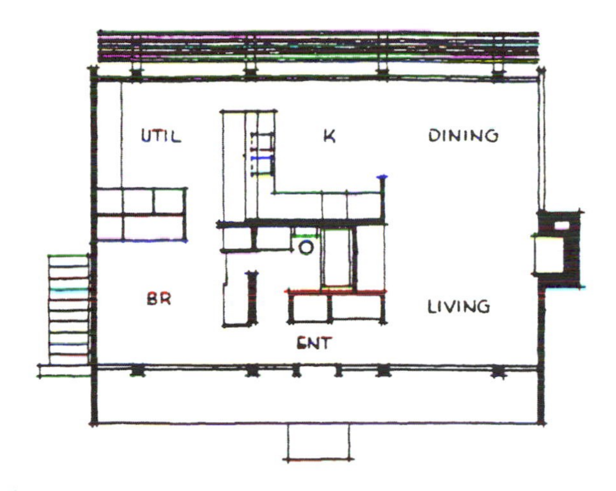

Figure 2 Simple structure around a central mechanical core: Post Structure Compact House – Paul Hayden Kirk, architect (Ford & Creighton 1954)

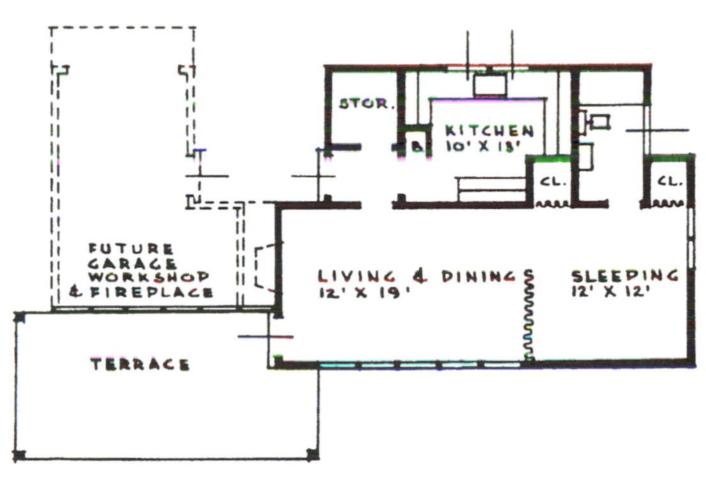

Figure 3 The large glazed wall creates a feeling of expansiveness in a space that is economical and compact: Houses for Homemakers – Royal Barry Wills, architect (Wills 1945).

easily be made and unmade through the extensive use of sliding walls and movable partitions. The principal problem beyond the accommodation of so many functions in such a small area was to design for a feeling of spaciousness. Homeowners in the postwar period did not (and still do not) want to be reminded that they lived in a small house. One strategy used to create the illusion of greater space was to emphasize the relation of the house to its immediate surroundings. This was achieved through the use of large plate-glass windows and glass patio doors which dissolved the confining feeling of conventional walls while expanding the perceived limit of the space outside the house. Another strategy was the use of drapery, accordion walls, or ceiling-high movable storage shelves as partitions. These had the effect of blurring the limits of rooms and of making them seem larger. The Houses for Homemakers project in New York State, built in 1945 and designed by Royal Barry Wills (figure 3), illustrates the opening up of an exterior wall to the outdoors to create a greater impression of size in a space that was both economical ($3,000) and small (85.5m² / 855 ft²). The multi-purpose room, fitted with a large glazed wall, was oriented according to the most

Figure 4 Efficient use of space in a simplified design: Place House – Roy Worden and Associates, architects (*Architectural Forum* 1949b)

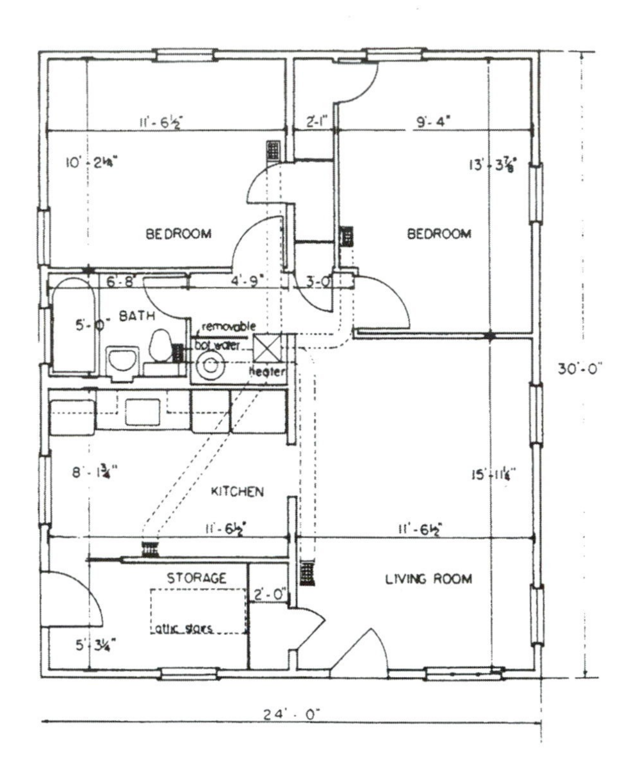

desirable exposure. By creating a living space that looked outward rather than inward, a smaller space could be utilized without compromising perceived spaciousness. The use of a sliding partition to set apart the sleeping area also validated the smaller overall living space, since the sleeping area could be tidied away and turned into a second living room when a larger space was required. The provision of a covered terrace was an example of indoor-outdoor living that allowed expansion of the inhabited area of the house without the extra expense of additional enclosed space. The designer also included the location of a fireplace and garage/workshop as potential later additions to the original plan (Wills 1945).

The Place House (figure 4), built in 1949 in South Bend, Indiana, and designed by Roy Worden and Associates, was also very compact (72m² / 720 ft²) and cost only $6,350. Efficient and simple allocation of rooms within a rectangular floor plan made the most of the available space. The low price was achieved through economic construction practices such as the use of precut framing members and prefabricated plumbing stacks. The utilities were moved into the foundation with a specially designed hot air system: a down-delivery furnace circulated heat through under-slab ducts to every room in the house. A storage room was provided, with additional storage space in the attic accessed by a retractable staircase (*Architectural Forum* 1949b).

On a more contemporary note, San Francisco architect Donald MacDonald's "garden cottage" is a compact prototype of 80m² (800 ft²) on two storeys. The design was used during the 1980s in a number of infill projects either as detached units or in rows and sold for the low (for San Francisco) price of $150 per square foot. The ground floor contains one bedroom (two if a garage is not included) and a bathroom; the second floor is unpartitioned and contains a living and dining area, an L-shaped kitchen in one corner, and a fireplace (figure 5). The ceiling on the second floor is pitched, allowing for a sleeping or storage loft accessed by a ladder. The home was designed to increase property ownership among renters, while rejecting middle-class concepts of space and use. MacDonald's design illustrates his belief that owned housing is a priority that supersedes traditional notions of size and layout (Pastier 1988).

In addition to keeping the housing unit compact, one of the guiding principles for the designer of affordable housing is to avoid complexity. A simple box is almost always the least expensive solution, which is why vernacular buildings such as farmers' cottages have always been rudimentary, compact volumes with little irregularity of contour or variation in shape. Needless to say, these are not ingredients that lead to exciting or spectacular architecture. Popular housing has always tended to the modest and plain, but there is no reason that it should be unattractive. Decorating simple boxes is precisely what Georgian architects did so well by specifying a well-placed moulding around the entrance door or a carefully proportioned window.

The traditional language of architecture is rich. The use of vernacular domestic elements such as porches, balconies, bay windows, gables, and dormers continues to be a useful way of imparting a sense of identity to individual homes while maintaining a cohesive street facade (Rybczynski 1989). Equally important, any such architectural language should be comprehensible to the public (figure 6).

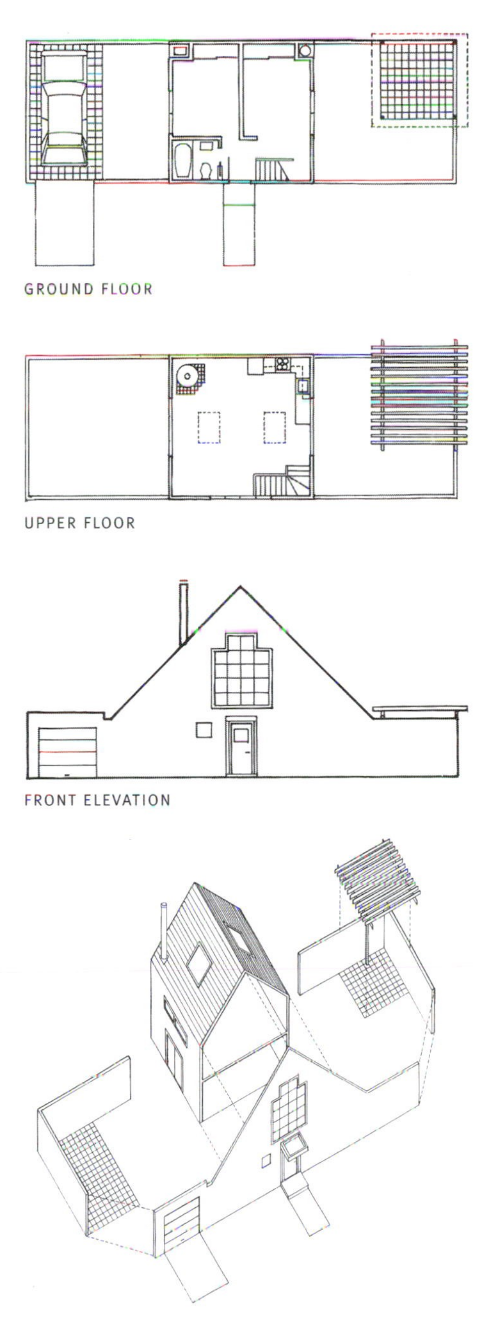

Figure 5 A two-storey, 20 x 20 foot box: Donald MacDonald's "garden cottage" (Pastier 1988)

Figure 6 Elevation studies of the Grow Home: an architectural language easily grasped by the public

Reducing the size of the pre-war house did not in itself provide an appropriate postwar living environment. Living space had to be re-evaluated in order to identify the essential spatial requirements of a small structure. The design strategy was straightforward: accommodate the greatest amount of functional space within a small, low-cost structure. An open floor plan was introduced to maximize the potential range of uses in the restricted interior space. This flexibility allowed the owner, not the designer, to define the space according to particular need: the owner could adapt the small house to a growing family's requirements for space and privacy. Designing for interior adaptability represented a significant departure from the preconceived floor plans of traditional design.

Designing for growth and adaptability is another cost-reduction strategy that allows occupants to modify their living spaces according to their evolving needs. A design strategy that permits the growth of the unit must, however, operate within numerous constraints. These affect fundamental elements of the design such as circulation, services, materials, light, and ventilation, any of which can stand in the way of change if the original design is insensitive to the fundamental principles necessary for growth. The initial design must therefore involve an approach that corresponds with a vast array of anticipated circumstances without compromising the objective of efficiency. Well established within the principles of flexible design, add-on and add-in strategies offer viable methods of fulfilling the future needs of a dynamic household.

A common approach to future expansion is the add-on method. This strategy involves the construction of additional rooms that can be are attached to the original structure as the need arises for extra space within the home. Particularly suited to households whose initial space requirements are minimal, this method allows expansion of the home at the discretion of the owners and as an inevitable consequence of the growth of the household. Add-on potential offers many opportunities to households, among them a smaller initial investment that enables people to become homeowners and feel confidence in their investment by securing the flexibility to expand the unit at a later date.

Tailoring a design to accommodate future additions presents several obstacles. Space must be available for any new construction, and will entail extra costs to the homeowner incorporated within the original investment. Further, restrictions involved in applying this system of expansion to housing types other than detached, ground-related dwellings will depend upon accessibility to exterior space. To improve the chances of success, all strategies must be carefully considered to ensure that they include provisions for future growth within the original design.

The second strategy for expansion – add-in – involves the provision of unfinished spaces within the original dwelling which can be finished at a later date as the need and means arise. In comparison with add-on procedures, this process requires a somewhat higher initial investment in space and in structure – compensated, in return, by a considerably smaller investment at the time of expansion. Nevertheless, it can prove to be an economical and efficient strategy through money saved on finishes and expensive labour. The expanding do-it-yourself industry provides the opportunity for additional savings by allowing homeowners to pay only for finishes within the areas they currently use, and then to pay only the cost of materials for the expansion area by reducing or eliminating private labour expenses.

The successful implementation of the add-in strategy is directly related to the livable quality of the space into which the homeowner will eventually move. Unfinished attics or basements can be a challenge in terms of light and ventilation. The unfinished space must also be redundant to the immediate needs of the household so that it does not interfere with overall occupancy while the space remains unfinished. A space that is essentially removed from the primary, prioritized rooms of the home will not cause problems that could render the home less attractive to buyers.

Flexibility is an essential part of any strategy for growth and adaptability. The procedures involved in the transition phase from the original unit to a larger living area can be significantly affected by the choice of materials used in construction. Concrete block and masonry may limit the flexibility of the housing unit because of their permanent nature. On the other hand, gypsum walls and prefabricated partition systems designed to be dismantled easily can make changes to the unit simple. Expenses can also be reduced by incorporating a modular dimension within the anticipated area of expansion, with additional components from the same system as the original structure, thereby cutting down on labour-intensive procedures.

The direction of growth is another key element of the strategy of designing for add-on or add-in (figure 7). Expansion must also be considered in the wider realm of the urban context, and a strategy should be employed that best harmonizes the original unit and its surroundings.

In an appropriately planned home, living space can be increased with minimal disturbance to the existing structure. The 110m^2 (1,100 ft^2) bungalow designed by Gill and Bianculli in Tennessee in 1946 (figure 8) allowed for expansion on the exterior according to a predetermined plan. The designers arranged the circulation so as to minimize disturbance during expansion. Furthermore, the bathroom was centralized so that it

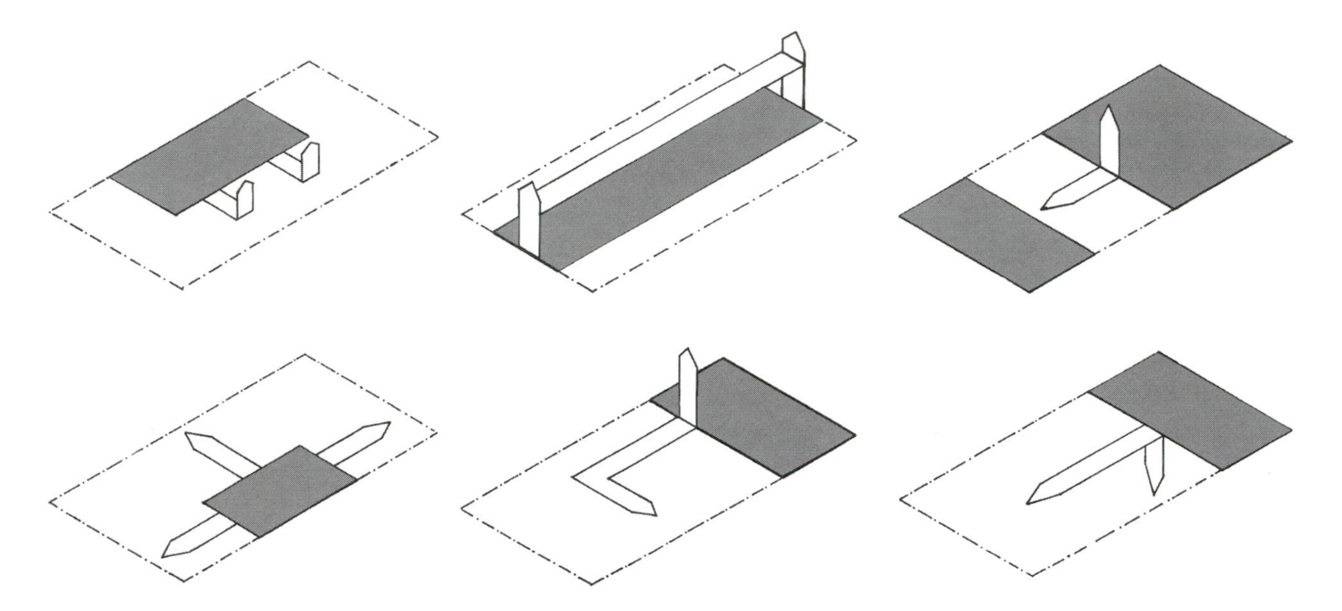

Figure 7 Direction of growth based on add-on and add-in strategies (Friedman 1982)

could conveniently serve additional bedrooms. Heating was also located centrally, reducing the amount of duct-work needed to heat the house. Initial construction costs were $4,349, with an estimated cost of $1,500 for the addition (*Architectural Forum* 1946a).

The Expandable House, designed by Robert Elkinton in 1954 in Mississippi, was a compact (86.8m² / 868 ft²), low-cost ($6,000) home planned and built for future expansion. The key here was to use a limited number of fixed features. The original plan called for two bedrooms, living and dining rooms, and a kitchen. The plan for expansion included the conversion of the living room into a third bedroom, the extension of the kitchen into the dining room, and the construction of new living and dining rooms (figure 9). The original version called for a double fireplace even though initially only one would be used, in order to allow for expansion without disturbance to the house's masonry.

The Chatham Houses, a single-family-unit project of forty semi-detached dwellings in England by the Chatham Borough Architecture Department, involved a design of single-storey, L-shaped units with garages and open yards in the front and an enclosed yard in the rear. At the initial stage the owners had the option of purchasing a basic one-bedroom unit, even without the garage if they so wished. The original unit of 65m² (650 ft²) had the potential for expansion

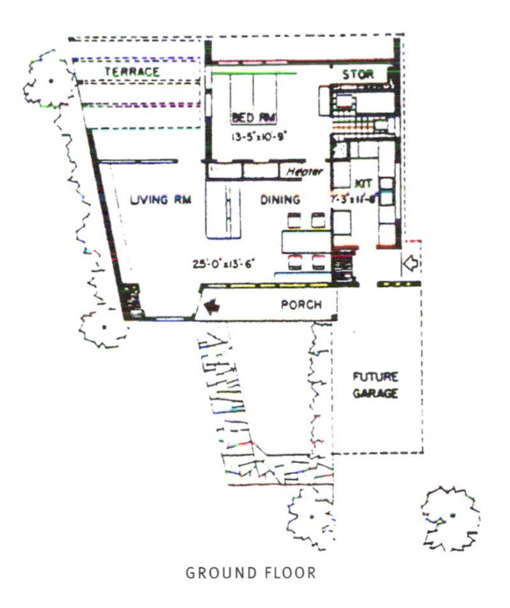

GROUND FLOOR

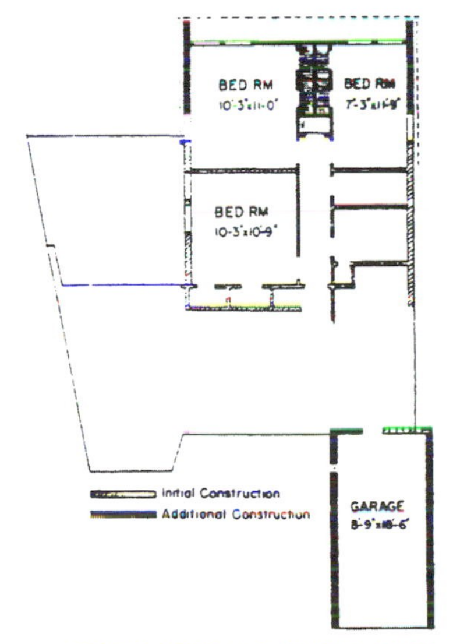

SECOND FLOOR IN EXPANDED HOUSE

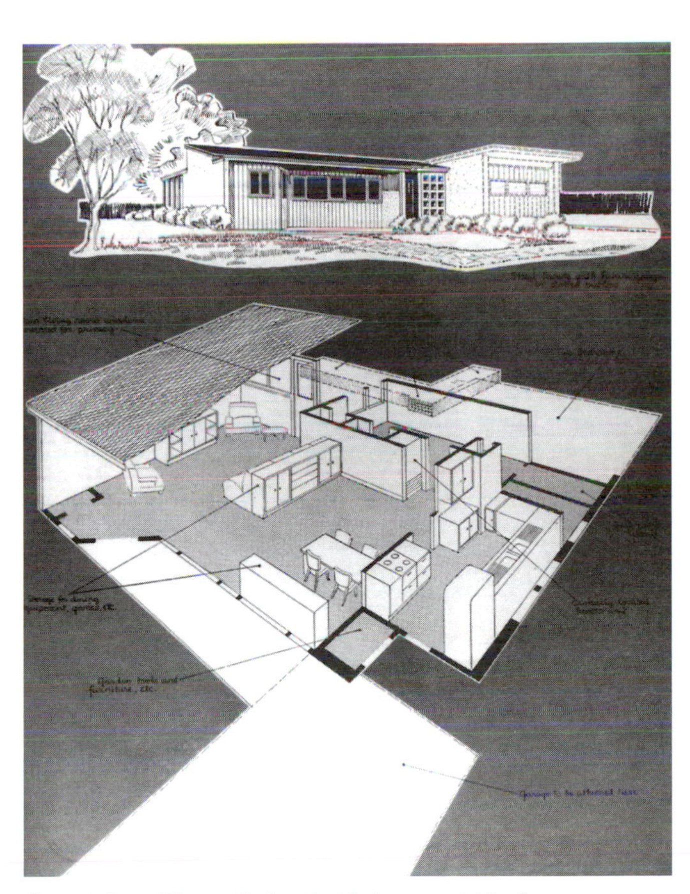

Figure 8 A small house designed with the potential for future expansion: the Expandable Bungalow – Gill and Bianculli, architects (*Architectural Forum* 1946a)

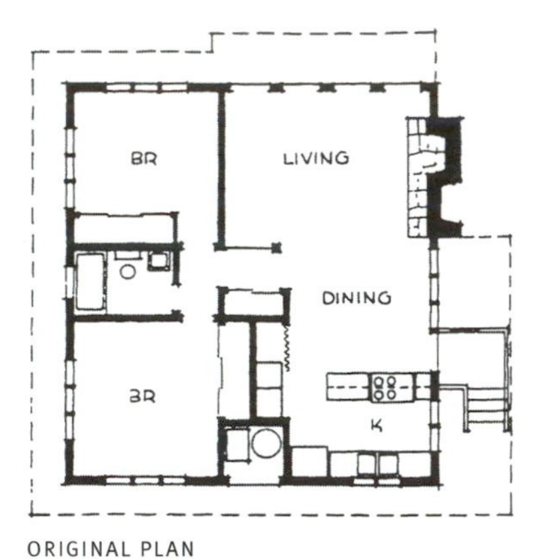

ORIGINAL PLAN

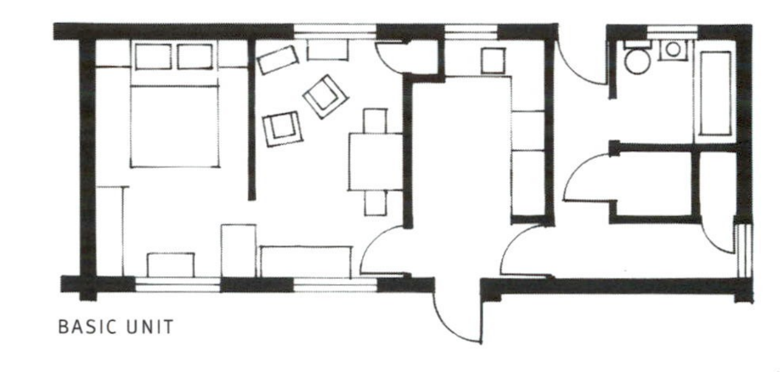

BASIC UNIT

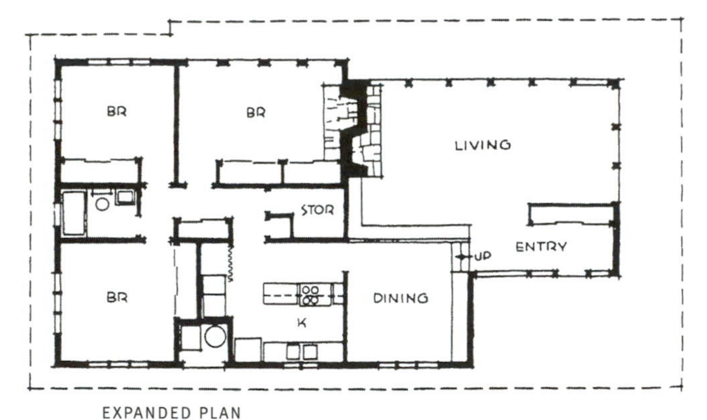

EXPANDED PLAN

Figure 9 A low-cost home which would allow for future expansion: Expandable House – Robert Elkinton, architect (Ford & Creighton 1954)

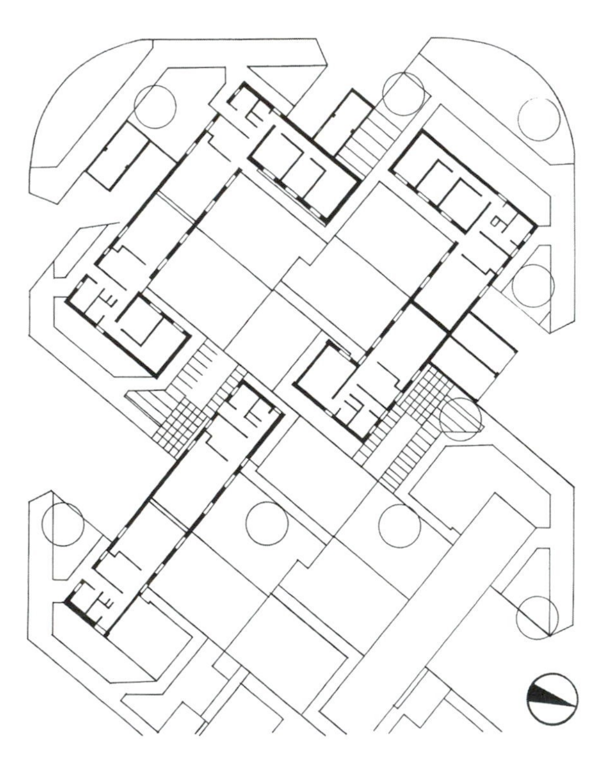

THE BASIC UNIT AND ONE, TWO AND THREE BEDROOM ADDITIONS

Figure 10 Community designed for horizontal expansion: Chatham Houses, England – Chatham Borough Architecture Department (*Architects' Journal* 1975)

into a three-bedroom unit of 85m² (850 ft²). The distinct L-shape of the building enabled the owners to expand the home in a horizontal, add-on fashion to the wing in the rear, and also in front if the garage was not an initial component of the unit (figure 10).

Design for growth was a predominant motif in the development of Ramat Magshimim, a cooperative agricultural village in Israel completed in 1978 by architect David Best. The principles and concepts

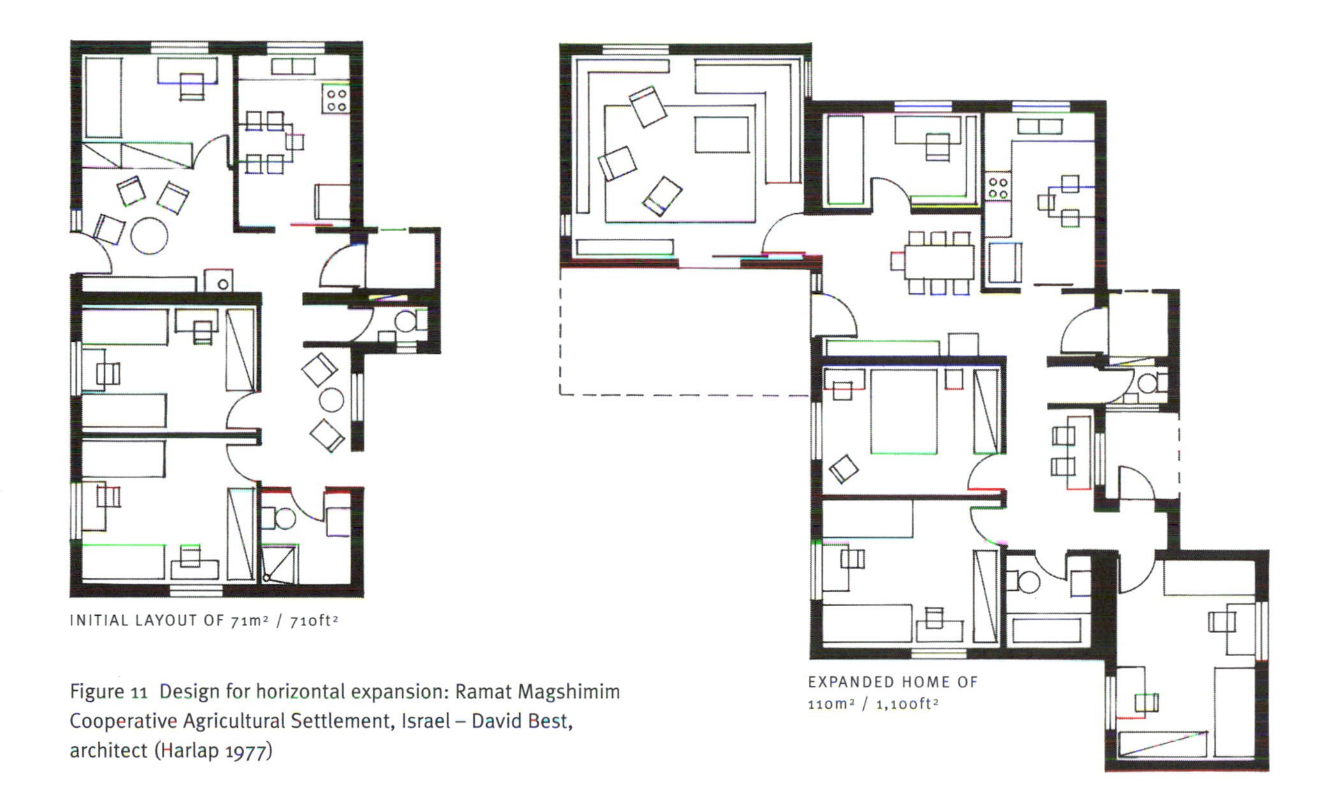

Figure 11 Design for horizontal expansion: Ramat Magshimim Cooperative Agricultural Settlement, Israel – David Best, architect (Harlap 1977)

of add-on expansion were applied not only to the homes but extended to include the entire settlement, village centre, and groups of houses. Flexibility was also employed throughout the development of 120 detached low-rise homes to allow for modifications during and after construction in accordance with the changing program and the specific requirements of the inhabitants. The planning of the units involved an organic hierarchy of open spaces and circulation which permitted expansion in the form of annexes to the units. The format of the original configuration was a 71m² (710 ft²) home marketed to young couples and singles (figure 11). According to the plan, rooms can be added gradually as the household grows until the final dimensional limits are achieved, a boundary that constitutes five to six rooms or 100 to 130m² (1,000 to 1,300 ft²). The external living spaces are defined primarily by the subsequent form of the additions themselves. With the incorporation of the potential for growth and adaptability into the original intentions of the scheme, the resulting development responds efficiently to the needs of changing households.

One of the simplest ways to expand living space is the internal – add-in – conversion of an enclosed, unfinished space. Vertical expansion is a popular

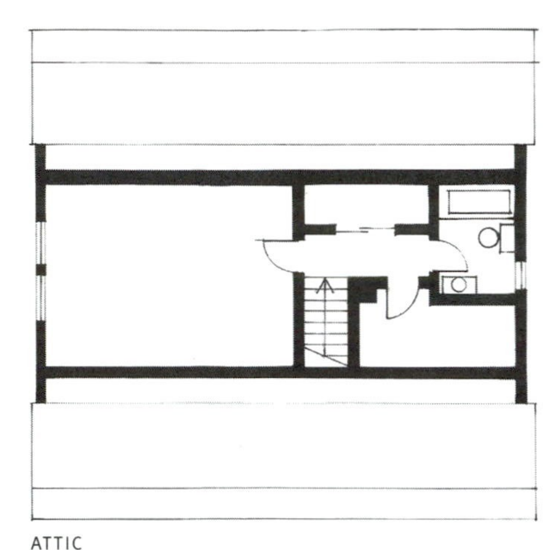

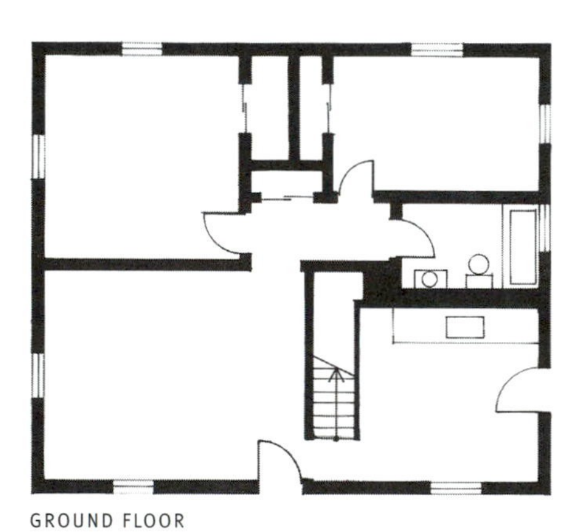

Figure 12 A mass-produced home based on the Cape Cod design with the potential for expansion to the attic: the Expansion Attic – Alfred S. Levitt, architect (*House and Home* 1954)

method: finishing attic space. In the Alfred S. Levitt Cape Cod Cottage (figure 12), the original plan of the house provided 34.5m^2 (345 ft^2) in the attic for storage or the later addition of a bedroom, storage space, and extra bathroom. Incorporated into the cost of $7,990 were appliances, a carport, and a fully landscaped lot (*House and Home* 1954). Conversion of garage space into enlarged living rooms or additional bedrooms is another economical method of add-in expansion.

The add-in strategy can be modified to fit the needs of many types of contemporary buyers: people who work at home need home offices, people with small children need nurseries, people with elderly parents need a guest room, and childless couples generally prefer large, open spaces. Different requirements generate different solutions in a small, inexpensive house since they lead to very different trade-offs. One way to accommodate a variety of trade-offs is to leave part of the house unpartitioned either on an upper floor – in effect, a loft space – or in the basement. Provision can be made for a service hook-up for a future bathroom. The homeowner may choose to use the upper floor or basement as a single open space or to build (or have built) partitions as required. Moreover, the unfinished space creates an adaptable and responsive home which can be completed according to the wishes, needs, and financial resources of the owner.

Another important element in the provision of affordable and flexible housing is the inclusion of the householder's own labour in the modification of the home. Unlike the North American custom of buying a new home with built-in cabinetry, the tradition in Europe is to provide homes without built-in kitchens and closets – components that are later installed by the homeowner. This strategy has several advantages. The quality of these elements – which can vary considerably – is left to the discretion of the homeowner. The occupants can choose a less expensive kitchen and upgrade it later or install the bare minimum of more expensive cabinetry and complete the job when funds become available. People who are handy with carpentry and installation can do the work themselves or with the help of friends. A distinct advantage to leaving the exact location of the clothes closets to the homeowner is that storage elements are placed only where required, making space use more efficient. Finally, the deduction of the cost of finished cabinetry from the construction cost of the home – and from the cost of the mortgage – can yield a substantial cash benefit. Every dollar included in the house mortgage translates into two or three dollars spent by the mortgage holder. Home improvements made through personal savings do not impose such a penalty.

The provision of kitchens and closets by the homeowner (not the builder) is made possible by a variety of furniture stores that sell unfinished, disassembled, and partially assembled furniture. Such stores sell not only domestic furniture but also sectional closets, cupboards, bathroom cabinets and fixtures, and modular kitchen cabinets and counters (figure 13).

Numerous cost-reduction strategies are available to the builder and designer of affordable housing. The building materials industry, for example, offers a wide range of product choices. Careful study, however, is required to identify and select materials and components that are environmentally sound and energy efficient and perform well over the life of the building. Using low-quality windows, for example, or reducing insulation values are poor strategies for saving money in the long run since the cost of future upgrading will be high. But significant savings can be realized through the use of less expensive finishing materials which account for a large proportion of construction cost where they can be easily and practically replaced with higher quality materials in the future. Vinyl tile, for example, is the least expensive flooring, but it can be replaced later with carpeting, wood parquet, or ceramic tile. The same principle applies to hollow-core doors, light fixtures, and interior paint.

Industrialized materials such as plywood, gypsum wallboard, asphalt shingles, polyethylene vapour barriers, board and batt insulation, and wall-cladding systems continue to revolutionize the building industry. Prefabricated elements already account for a large portion of the house: roof trusses, space joists, windows, pre-hung doors, and kitchen cabinets are all manufactured in factories. The large-

Figure 13 Do-it-yourself instructions for kitchen cabinet assembly. Top: Drilling through the frames to connect them with screws and sleeves. Bottom: Fine adjustments can be made in the vertical panels.

scale modular prefabrication of entire building components such as walls, floors, or even houses is less common. In North America panelized building – in which all walls are manufactured in a factory (with floors, roof and interiors finished on the building site) – is the more common prefabrication technology (figure 14). The modular manufacture of houses – as two or more completely pre-finished boxes – is the second most common prefabrication technology. Neither method, however, has managed to make a significant breakthrough in reducing construction costs (Wiedemann et al. 1989).

The success of large-scale housing prefabrication in Japan (steel, wood, plastic) and Sweden (wood) suggests that breakthroughs may yet occur in Canada. The Japanese experience indicates that prefabrication has the greatest chance for success when it is part of a comprehensive approach to housing delivery that integrates marketing, customer relations, design, manufacturing, erection, and post-occupancy servicing in one organization.

The prefabrication and mass production of housing have been inescapable elements of the affordable housing industry since the postwar period. As architect Royal Barry Wills predicted at the end of the war, the era forced designers to re-examine the relationship between cost and size (*Architectural Forum* 1945). Great faith was placed in technological innovations, particularly prefabrication, as a solution to the housing crisis. Although prefabrication had existed before the postwar housing boom, to the industry at the time it represented salvation. Several housing manufacturers successfully produced completely prefabricated homes, but many found it a cumbersome

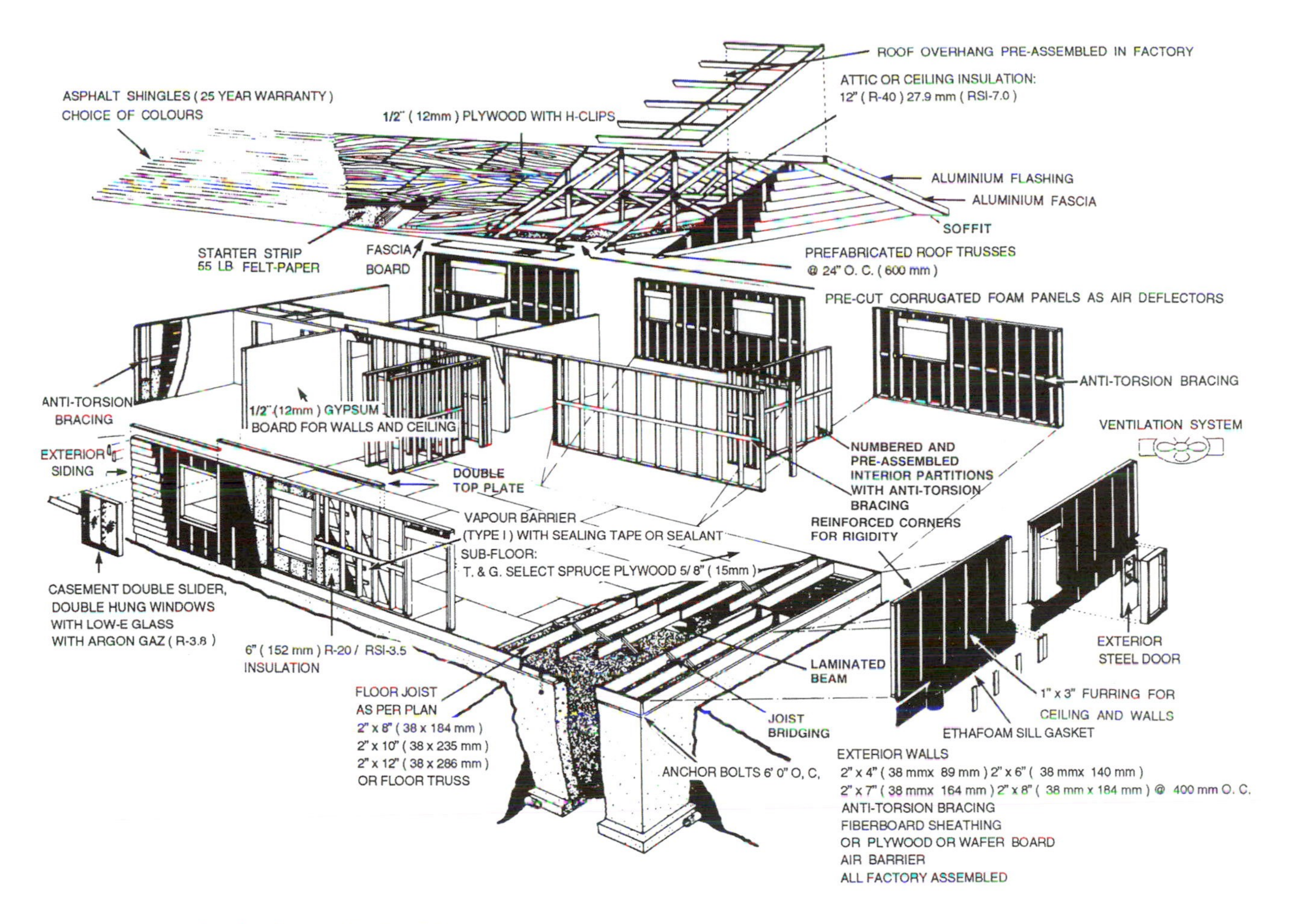

Figure 14 Perspective showing panelized prefabricated components by Modulex Inc.

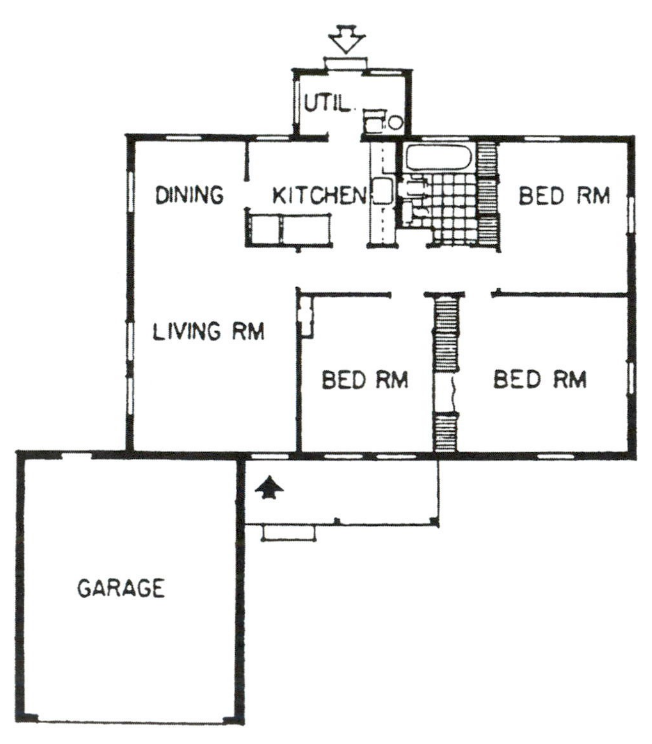

Figure 15 A prefabricated panelized shell allowed for one of five different styles: Kaiser Prefabricated Home – Henry Kaiser and Fritz Burns, builders (*Architectural Forum* 1946b)

and expensive proposition. It was more lucrative to prefabricate housing components rather than the entire house, as demonstrated by the prefabricated kitchen-bathroom unit. These two rooms were placed back-to-back and manufactured as an optimally cost-effective, single unit. Significant cost reductions were also achieved through the standardization of stock parts such as windows, doors, and wall panels, so that the feasibility of prefabricating these components increased considerably.

By 1951 one-fifth of an average house was made of prefabricated components (*Architectural Forum* 1951). In the Kaiser Prefabricated Homes (figure 15), mass production was the key to reducing construction costs. Henry J. Kaiser was a well-known industrialist and advocate of mass production who teamed up with the established developer Fritz Burns to produce this panelized home-building scheme. Factory production included full-wall, floor, and ceiling panels, plumbing equipment, kitchen cabinets, and a two-car garage. To avoid monotony, the houses were varied in appearance; site crews applied one of five "dressings": Cape Cod, Colonial, Ranch House, California, and Contemporary. The 90m² (900 ft²) homes could be purchased for $6,950 to $8,650 (*Architectural Forum* 1946b).

Mass-produced housing represented the culmination of housing innovations from many domains. The houses were designed to be compact and efficient. Materials and construction techniques were chosen to maximize affordability and the rate of production. Marketing strategies, including advertising by developers, were refined to encourage people to believe that mass-produced housing was the optimal housing option. But the downside of uncontrolled and poorly designed mass production of housing was the creation of instant communities made up of virtually identical, low-cost, small houses. The preponderance of such middle-class communities inspired the National Housing Authority to voice its concern: "The rapid acceleration in home building poses the risk of unplanned or poorly planned communities with insufficient attention being paid to such necessary facilities as schools, hospitals, utilities, stores, transportation and employment" (Bernardi 1946).

Writing about the modern suburbs in 1961, Lewis Mumford (p. 486) characterized the situation in more dramatic terms: "In the mass movement into suburban areas a new kind of community was produced, which caricatured both the historic city and the archetypical suburban refuge: a multitude of uniform, unidentifiable houses, lined up inflexibly, at uniform distances, on uniform roads, in a treeless communal waste, inhabited by people of the same class, the same income, the same age group, witnessing the same television performances, eating the same tasteless pre-fabricated foods, from the same freezers, conforming in every outward and inward respect to a common mold, manufactured in the central metropolis."

The advent of mass-produced housing introduced assembly-line housing construction. The introduction of power tools and labour specialization at the construction site is attributed to Alfred Levitt. Self-described as the "General Motors of the housing industry," Levitt and Sons divided house construction into twenty-six steps that incorporated the extensive use of prefabricated components. They produced their own nails and concrete blocks, they owned lumber mills and distributing firms for electrical appliances, and they strove to maximize the efficiency of the house construction process. The result of this carefully controlled assembly process was an 80m² (800 ft²) house completed every fifteen minutes. The early Levitt houses came in two styles: Cape Cod (figure 16) and Ranch House. At a cost of $7,900, they included appliances, a fireplace, plumbing, and wiring (*Architectural Forum* 1950a).

Despite their uniform appearance, the Cape Cod Cottages and Ranch Houses produced by the Levitts were very popular. At the Long Island development known as Levittown, people stood in line to buy "one huge mass of homogeneous mass-produced housing, all pre-planned to standardization and mass production" (Gans 1967). The popularity of these homes was due largely to their economical attractiveness to returning veterans. Using federally insured loans administered by the Federal Housing Authority, they could obtain thirty-year mortgages with no down payment. This meant that in 1950 payments for mass-produced homes could be made at $56 per month. At the time the average monthly rent for an apartment in the city was $93 (ibid.).

Herbert J. Gans, writing in 1967 on the Levittown built in New Jersey between 1955 and 1959, describes the lessons Levitt had learned in the two previous towns built by the firm in New York and Pennsylvania. The New Jersey community (its name was later changed to Willingboro) offered the innovation of mixed housing types (Cape Cod, Rancher, and Colonial, ranging in price from $11,500 to $14,500) to deflect the criticism of city planners who spoke out

Figure 16 Mass-produced, prefabricated house with all the mod cons, including sliding aluminum windows and built-in television: Levitt House – Levitt and Sons, builders (*Architectural Forum* 1950a)

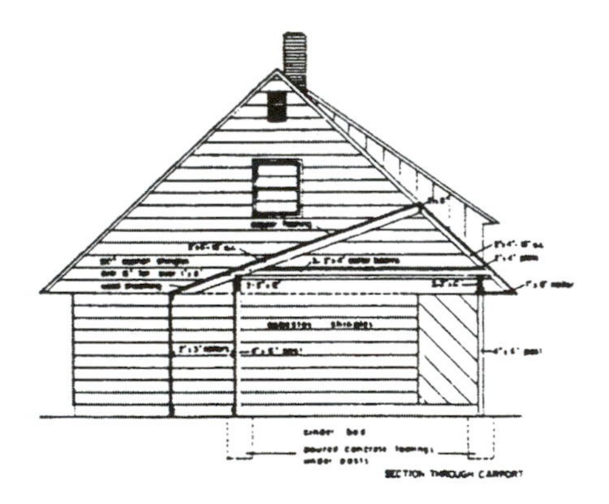

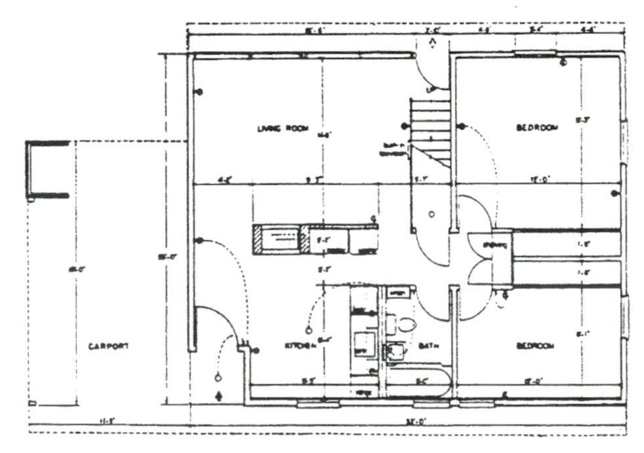

against the homogeneity of postwar suburban subdivisions. Levitt had built the New York development for lower middle class returning veterans and the Pennsylvania suburb for blue collar and "marginal" buyers who could afford the homes only because they did not need to provide a down payment; the New Jersey Levittown was aimed at the middle/upper middle class in order to increase the company's profit margin. The Colonial home design was promoted to attract higher-income purchasers who, the Levitts suspected, were ready to reject the abstract and severe contemporary designs which had dominated the architectural magazines of the late 1950s. The assumption made was that family living was the main priority and that preferences of style came second (Wright 1981). For Levitt the primary design considerations were livability and the effective use of space. New house models were improved versions of earlier models that had been altered in response to market demand, technological innovation, and cost-cutting opportunities. The basic design included a concrete slab, precut materials put together on an assembly-line basis, and an open house plan with maximum space for rooms and storage and minimal space for circulation and foyers. The Levittown concept was aimed at young families buying their first home, and as such the houses were designed with small children in mind: the bedrooms were large enough to serve as playrooms, an extra bathroom was provided for children, and the kitchen was located so that the mother could watch youngsters playing outside (ibid.).

Whether a house is attached or detached is another key element in reducing construction cost. The house that shares its walls with its neighbours – described variously as a rowhouse, townhouse, or terraced house – is a traditional urban housing solution that dates back to at least medieval times (figure 17). Its chief advantages are that it can be built on a fairly narrow plot, typically between 6.1m and 7.3m (20 and 24 feet) wide, which allows relatively high densities but which also incorporates most of the advantages of the detached, single-family home: a private front door, easy access to the ground, and a clear definition of a public street side and a private rear

Figure 17 Traditional rowhouses in Amsterdam

garden. The narrow width has a profound effect on cost: in a time of accelerating land and infrastructure costs, it is the single most effective cost-reduction strategy. The chief constraint on the rowhouse is the width between the shared walls on both sides. With only two facades available for windows, width governs depth as well as the number of rooms that can be placed against the two exterior, windowed walls at the front and rear.

In general, societies have tended towards the efficiency of dense housing types such as rowhouses for the purposes of social interaction, defence, shared resources and facilities, transportation, and tradition (Van der Ryn and Calthorpe 1986). In medieval England, where a high value was placed on trading street frontage, narrow and deep plots often had a ratio of width-to-depth in excess of one to six. In Chester, a medieval city built on Roman ruins, merchant houses called The Rows consisted of a shop in front, a hall and courtyard in the middle, and a kitchen in back, all linked by a long side passage; bed chambers connected by a gallery occupied the upper level (Schoenauer 1992). During the Industrial Revolution the rowhouse became the chief housing form in cities both in Britain and America. In nineteenth-century London, houses (i.e., rowhouses) were classified according to width: Class I was 6.1m (20 feet) wide and four floors high; Class II was 5.5m (18 feet) wide, shallower than Class I and three floors high; Class III was 4.9m (16 feet) wide, two floors high, with a basement; and Class IV was 4.6m (15 feet) wide, two floors high, without a basement (Muthesius 1982). The important distinction was between the first two classes, which were wide enough to be subdivided into two rooms, and the second two classes, which could accommodate only a single room across the width of the rowhouse.

The narrow rowhouse has been the subject of study for many modern architects interested in reducing construction cost. An early example was designed by J.J.P. Oud for the 1927 Weissenhof Exhibition in Stuttgart (figure 18). The width was 5.5m (18 feet), which permitted the bay to be subdivided into one narrow and one wide bedroom. The “Roq” and “Rob” housing, designed (though unbuilt) by Le Corbusier for Cap Martin in 1949, proposed a 4.6m (15 foot) wide rowhouse, a dimension that allowed two extremely narrow bedrooms side by side. The settlement of Siedlung Halen, a version of this project, designed by Atelier 5, was built on the outskirts of Bern in 1961. The Halen community consists of eighty-one rowhouses 4.9m (16 feet) wide, arranged in two staggered rows in a picturesque woodland setting. The project was regarded as an alternative to suburban housing (considered to be sparse and uninteresting) and to high-rise urban living (deemed unsuited to families). Siedlung Halen has long been designated as a model of high-density, low-rise housing since it offers dense homes in a communal setting without sacrificing individual privacy. As a link with tradition, the community is a modern interpretation of a standard housing design of medieval Bern: an urban Swiss building form that occupies a long, narrow slot of space (Sherwood 1978).

In Canada, LeBreton Flats, a three-hundred unit development initiated by Canada Mortgage and Housing Corporation and undertaken in the early 1970s in Ottawa on a site originally settled as a lumbering community, is a narrow-front development

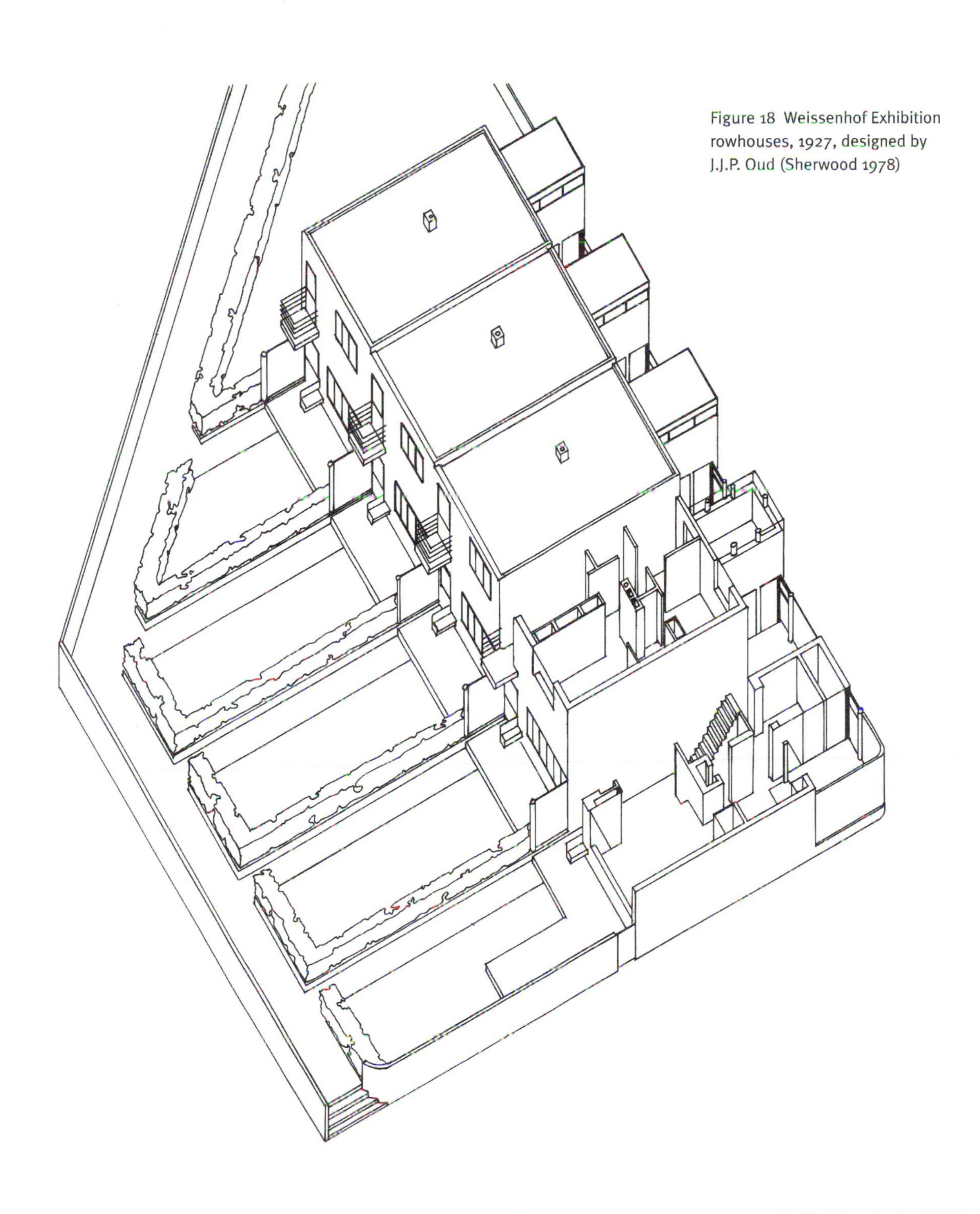

Figure 18 Weissenhof Exhibition rowhouses, 1927, designed by J.J.P. Oud (Sherwood 1978)

designed by Ian Johns. Land parcels were sold to private builders and non-profit co-operatives, with co-ops allowed input into the design process. Three-bedroom units, some with living rooms one and a half storeys tall, sold initially for about $65,000. Every unit had a garage in front, a large outdoor area, and street frontage. The multi-level style gave rise to a townhouse plan that became popular with developers in Ottawa, Calgary, and Vancouver. In Ottawa's Lowertown, Cathcart Mews, designed by Johns and completed in 1981, featured sixty-three units at 4.9m (16 feet) wide. Residents were satisfied with the openness of the design, which offered a greater sense of space than the actual 100m^2 (1,000 ft^2). Screens, terraces, and gardens provided necessary privacy. Site planning and the positioning of each unit were design priorities for Johns – more important than the narrow width, which at no point became an issue, either at the design stage or for city zoning.

The narrow-front rowhouse, which can be as narrow as 3.7m (12 feet), can successfully accommodate a small family. On the ground floor, 4.3m (14 feet) is wide enough for a living room and a kitchen-dining room. Two generous bedrooms can fit on the upper floor, and a third bedroom or other functions could be added in the basement. The small structural span reduces construction cost since standard sized joists are used, and this allows flexibility in internal partitions by eliminating the need for interior load-bearing walls.

The narrow-front rowhouse has important implications for land use. A one-storey bungalow on a 18.3 by 30.5m (60 by 100-foot) lot yields a gross density of approximately twelve homes per hectare (five per acre). A two-storey, 4.3m (14-foot) wide rowhouse on a 4.3 by 30.5m (14 by 100-foot) lot yields a gross density of sixty homes per hectare (twenty-four per acre). Expressed another way, a hectare of land can accommodate about fifty people (twenty people per acre) living in bungalows. The same amount of land – and the same amount of roads, sewers, water lines, and storm drains – can accommodate more than four times as many people living in narrow-front rowhouses. Regarded in this most basic manner, narrow-front housing is more than four times cheaper than single-family, detached housing.

The effect on the environment of denser neighbourhoods is considerable: less automobile travel, fewer roadways, less runoff – in general, less energy and expense invested in community infrastructure. Rowhouses provide significant savings in building materials and energy consumption because of the shared walls. A two-storey narrow-front rowhouse has only one-third the exterior area of walls and one half the roof area of a similar sized bungalow; heating and air-conditioning costs are accordingly lower.

Rowhouses are a flexible housing type that, depending on a number of constraints, can produce a variety of different configurations: simple street rows, pairs of semi-detached houses, terraces, and courts. The constraints vary according to a number of factors: densities, parking solutions, forms of ownership (freehold, condominium, cooperative, or rental), relationships to the street, and uses of public space. The way in which rowhouses are laid out is a function of the desired density, which is itself the result of land prices and urban character. A suitable solution for suburban developments would be groups of houses formed around landscaped public gardens. Semi-detached rowhouses provide many of the benefits of individual bungalows, at a considerably higher density. For example, a rowhouse can be built on a 6.1m (20-foot) wide plot, sharing a driveway with the neighbouring house. This produces a net density of fifty-two homes per hectare (twenty-one per acre). Straightforward rows, with or without back lanes, produce a more urban environment with a higher net density. The highest density solution is the housing terrace which was common in Montreal at the beginning of the twentieth century: the homes have a shared, landscaped garden in front and a service lane that provides access to a basement garage in the rear. There is no back garden. The housing terrace yields a net density of eighty homes per hectare (thirty-two per acre) and is suitable for inner-city neighbourhoods.

The need for parking distinguishes the modern rowhouse from its ancestors. The preferred solution is to provide parking spaces on the individual lot, either in the form of a basement garage or at the rear of the garden, accessed by a back lane. The back lane is a traditional solution that offers many advantages since it can also function as a service access for garbage pick-up, oil delivery, and so on. Parking at the front of the house can be on the street, on a paved area in front of the house, or in a basement garage. The disadvantage of all three front options is the effect they have on the appearance of the street. The third alternative is to park in small, common lots. This is an efficient solution with regard to use of space, although it is usually the least popular with the individual homeowner for reasons of convenience and security. Not much can be done about the convenience aspect, since group parking always requires a short walk from the lot to the house, but security can be improved by providing enclosed parking areas with individual lockable garages.

The form of home-ownership affects the configuration of the housing. Freehold requires a clear demarcation between privately owned land and the street. Rental housing, co-operatives, and condominiums are more flexible, since it is possible to maintain land for the common use of the tenants, co-op members, or owners. Common land can include gardens, pedestrian areas, back lanes (if they are not municipally owned), recreational facilities, and day-care centres.

Different types and arrangements of rowhouses create various relationships to the street. Simple rows of houses define the street in a traditional manner, forming a solid wall. Semi-detached homes create a looser environment, more like a traditional suburb. Housing courts or terraces provide landscaped spaces of a semi-public nature than can create the effect of small parks.

A well-designed neighbourhood would be a combination of different housing solutions to provide a variety of housing types and street forms. The narrow-front rowhouse is only one type of solution – one out of many. A neighbourhood should combine many housing forms (small apartment blocks, detached houses, flats over shops) that respond to people with different incomes and lifestyles. The urban patterns displayed in figure 19 integrate several housing forms in an attempt to avoid the uniformity and monotony that too often characterized planned developments in the postwar period.

Figure 19 Grow Home urban patterns

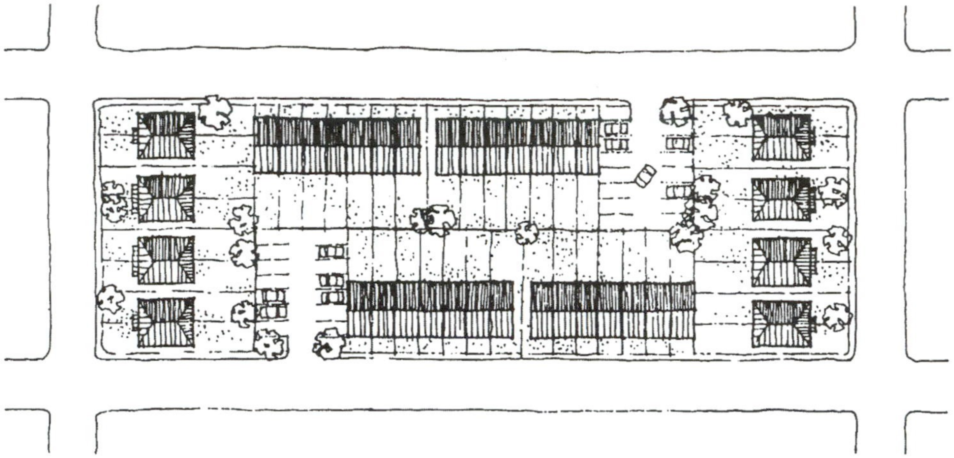

Mixture of semi-detached units and rowhouses (density of 60 units per hectare / 24 per acre)

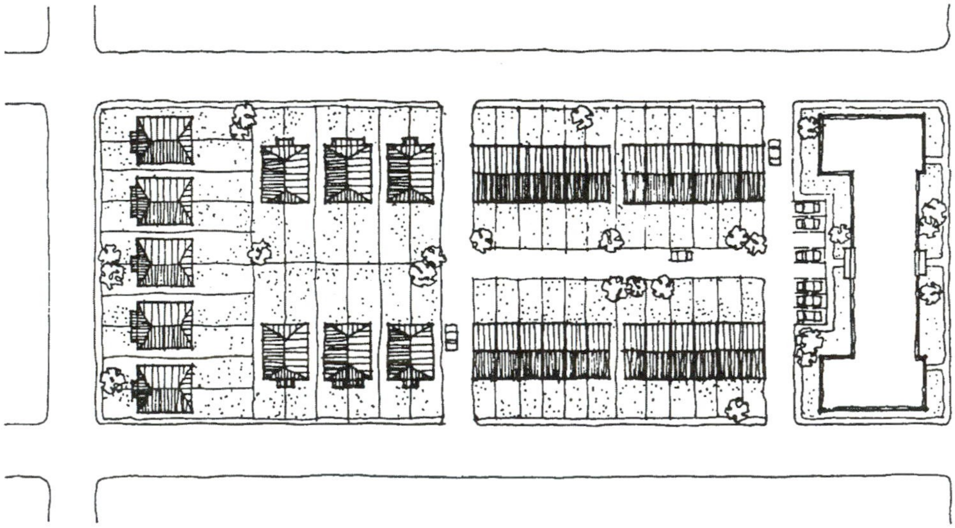

Mixture of semi-detached units and rowhouses served by a rear lane, with an apartment building at one end of the block (density of 65 units per hectare / 26 per acre)

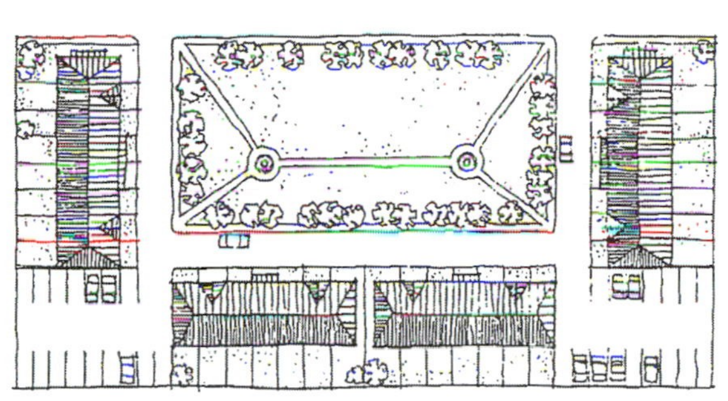

Housing terraces on three sides of a park with group parking in open corner lots or in private garages (density of 42 units per hectare / 17 per acre)

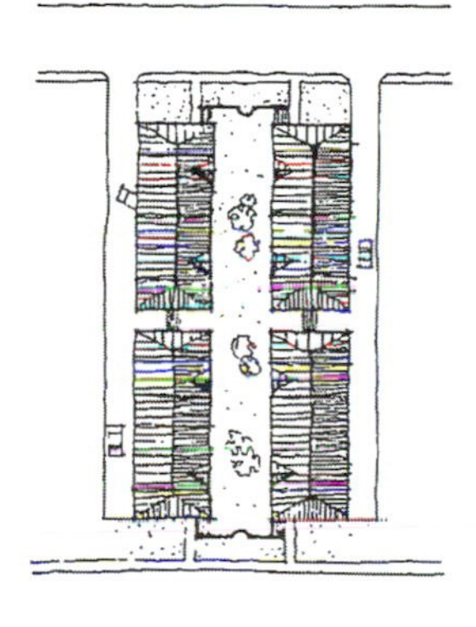

Housing court with individual entrances through a common garden and parking in basement garages accessed from a rear lane (density of 80 units per hectare / 32 per acre)

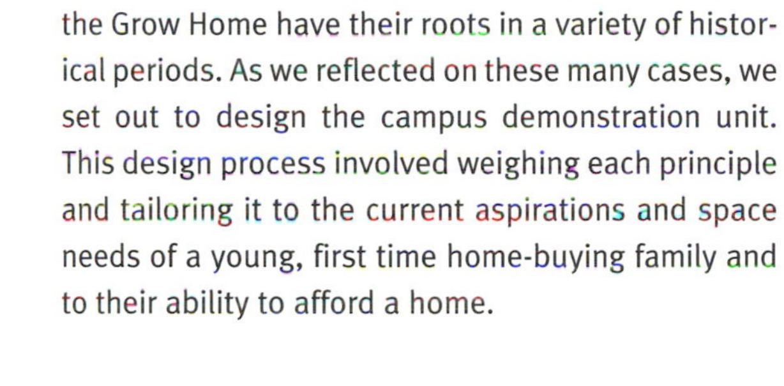

Many of the strategies that inspired the design of the Grow Home have their roots in a variety of historical periods. As we reflected on these many cases, we set out to design the campus demonstration unit. This design process involved weighing each principle and tailoring it to the current aspirations and space needs of a young, first time home-buying family and to their ability to afford a home.

3 THE PROTOTYPE

Most homes in North America are built by developers. Unlike custom-built homes, the design of such mass-produced homes demands a different process. At the outset the architect has only a few components to work with. The cost of a serviced lot in the development is known. The developer (builder) familiarizes the architect with built models that have been successful in the past. He also points out the features that are very popular in a particular market (large bathrooms, for instance, or an efficient kitchen), as well as a sketchy image of potential buyers (first-timers in their twenties, for instance, with a medium-size income and about to have their first child). The architect and the builder then establish the main features of the home (size, number of rooms, garage), and the design process begins. For a large site, several prototype models will be proposed. The builder reviews the drawings – he may even canvass the opinion of his real estate agent – and returns to the architect to suggest changes. A home will slowly turn into a product. It will find its way into the firm's marketing brochure and into the real estate section of the local paper.

The design of the Grow Home began with assumptions of the trade-offs that home buyers would be willing to make, what they would settle for, and what they really wanted. When the design phase finally arrived, we spread out sketch paper, pulled out rulers and scales, and sharpened our pencils. Of course, we needed a context. We needed a client (that is, a builder who would engage our services), we needed the builder's target market (*his* clients), and we needed a location where our designs could be built.

What was *our* context? What were *our* known pieces of information? Our home would be constructed by a small development firm. The site would be within the metropolitan Montreal area. The land price was $15 to $20 per square foot in the city, $5 to $10 in suburban towns.

The ultimate users would be small families, probably first-time buyers with or without kids, single people or single-parent families. The calculation of the selling price was based on the income of a young working couple or a single earner and led to the assumption that they would be able to afford a home with a price between $60,000 and $80,000 (including land). We knew that these buyers would have to make trade-offs, but what would they consider as absolutely essential in a new home? For starters, at least two bedrooms and space for a third (they would probably have a second child). They would also want a living room, dining area, study, and at least one bathroom but preferably two. We assumed that they might be willing to settle for smaller portions of these spaces. Now we were ready to design.

The sure way to reduce the cost of land and services is to build on smaller lots. The depth of typical subdivided lots in Montreal is 30.5m (100 feet). This was our starting dimension. What then should be the width? We determined this dimension by examining our own city, Montreal. The rowhouse is an integral part of the urban housing vocabulary here. The city is renowned for its plex building type with its exterior metal staircase. Montreal builders have long recognized the cost-saving advantages of joining buildings: among them, saving on facade construction and reducing heat loss. We knew that if we introduced a stacked and joined unit, buyers and builders would not raise an eyebrow. The main difference in our design would be that each vertical slice would house a single household rather than two or three as is the case with the classic plex.

What about the width? If our assumed buyers wanted two bedrooms, they would likely be located on the upper floor. (In a stacked unit, the public functions are usually on the main floor with the bedrooms located on the upper floor.) The dimensions of a decent-sized master bedroom are anywhere from 3.7 to 4.6m (12 to 15 feet) by 4.0 to 4.6m (13 to 15 feet). We also knew that in order to proceed economically in the building of a home, we would have to specify that standard construction components be used. If the floor joists were to be 50 by 250mm (2 inches by 10 inches), their standard available length is 4.6m (15 feet). Mass production had been a definite consideration for us. If a number of units were to be constructed on the same project or if they were to be built in inner cities where parking and time are constraints, prefabrication would be an advantage. According to transport regulations, the maximum width of a transportable prefabricated module that does not require escort is 4.3m (14 feet). We knew that our chosen width might only be a minimum (although one finds rowhouses in Europe that are only 3.7m [12 feet] wide). Builders – and buyers, for that matter – might choose to expand the width to 6.1m (20 feet) and have three bedrooms. But perhaps the greatest advantage had to do with cost reduction. With land costs of $200 per square metre ($20 per square foot), standard lot depths of 30.5m (100 feet), and servicing costs of $1,320 per linear metre ($400 per linear foot), every metre of frontage cost $7,920 ($2,400 per foot). (These and all subsequent figures are in 1990 dollars.) By reducing the frontage from 5.5m (18 feet) – the minimum for private ownership required by the City of Montreal – to 4.3m (14 feet), a saving of $9,600 per unit would be achieved. Co-operative or condominium arrangements could provide ways around this obstacle.

What would be the length of the unit? Our thought process was guided by several principles. In addition to the master bedroom, a second bedroom would be needed, one that could be slightly smaller than the master bedroom, measuring 3.1 or 3.7m (10 or 12 feet) by 4.3m (14 feet). A bathroom on the upper floor would be a must, as would be a clothing or linen closet. When we added up the numbers, we arrived at a house that would be 9.8 to 11.0m (32 to 36 feet) long. We then considered the functions that the ground floor of such a house would have to accommodate, and we managed to fit a kitchen, dining area, bathroom, and living room into a space 4.3m (14 feet) wide by 11.0m (36 feet) long. We realized that buyers would have to do without a formal dining room. This would be one of the trade-offs that they would have to make.

Another approach to determining the unit length came from an examination of the city's zoning bylaws. Cities in general require that homes be recessed from the sidewalk. Unlike European townhouses in old cities that were constructed right to the front edge of a lot, we had to respect a minimum distance known as a setback. A rear setback also exists. Devised as an antidote to the spread of diseases during the Industrial Revolution, the legislation of an open space in the rear of a house was intended for the provision of sun and fresh air to each dwelling unit. The allowable built area of a lot, after deducting from it the land that is allocated for front and rear setbacks, is called land coverage, that is, the amount of area that the building itself may cover on the entire lot. When we calculated the land coverage on our 4.3 by 30.5m (14 by 100 foot) lot, we arrived at a building with a length of 11.0m (36 feet), which coincided with our previous arithmetic.

We then reflected on another idea. If the residents of our home eventually found their place to be too small and wanted to expand, they would not be able to grow outwards (add-on) – the home was already designed to its maximum length. Building upwards would also not be a consideration since city bylaws limited the height to 10.4m (34 feet), the upper edge of our roof. Home buyers would have to expand *within* (add-in). But where exactly? We considered two options. The first was to leave the upper floor unpartitioned. We envisioned a space that would be finished yet clear of partitions, much like a loft. The second option was to add a basement to the house. A common feature in most Canadian homes, the basement would be left unpartitioned and unfinished at first – a place for laundry, storage, and mechanical services – but could later be converted to other uses. The advantage of a basement would be that while renovation took place below ground level, life could carry on as usual upstairs. If we raised the foundation somewhat and designed large basement windows, the dark underground feeling could be alleviated. In some cases, the basement could even become a garage if the owners – a single person or a childless couple – did not need any more space than that available on the other two floors. This garage alternative became part of a range of options that we proposed.

Builders usually design and construct a model unit since they do not know who will be the ultimate buyers of their homes. They use this showcase home to display options which are chosen and paid for by the buyers. Some of the options are included in the cost of a basic unit, but the price of others – known as "extras" – is tacked onto the base price. For example, the offered flooring of a living room might be carpet, but if buyers wish to have wood flooring they can pay the difference in cost as an additional amount. The list of options also includes alternative interior arrangements of the same unit or a range of components that can be added to the unit, such as a balcony, deck or porch. The builder decides on the options based on the project location and the architectural styles with which buyers in that location are familiar. The available selection also depends on the extra amount that clients are able and willing to pay. Sometimes builders will accommodate a buyer's special request even if it is not one of the official options on offer. Builders prefer to build identical units since this reduces time and supervision, but in order to sell a unit they will tolerate exceptions or an option for which they charge administrative fees in addition to profit. All these extras frequently make up a sizable portion of the builder's profit.

In our design we suggested a basic, no-frills model and a range of exterior and interior options of differing quality and type, each option with a price tag (figures 1 and 2). The options in figure 2 break down into three categories: architectural configuration, exterior finishing, and interior finishing. Architectural configuration includes a choice of foundations: slab-on-grade, basement, or basement garage; balconies, decks or porches; different roof styles (sloped or flat); roof dormer; upper level unpartitioned

Factory Costs	**$ (1990)**
Upper floor	1,119
Exterior walls & interior partitions	1,942
Roof trusses	730
Roofing	730
Vinyl siding	545
Insulation	1,045
Windows & exterior doors	2,575
Gypsum wallboard	767
Interior doors & moldings	432
Kitchen cabinets, bathroom vanity	1,751
Staircase	352
Loading onto truck	495
Transportation	660
Total factory costs	13,223
Foundation slab	4,132
On-site erection	8,000
Subcontracting	11,557
Total Construction Cost	**36,912**

Figure 1 Grow Home base model

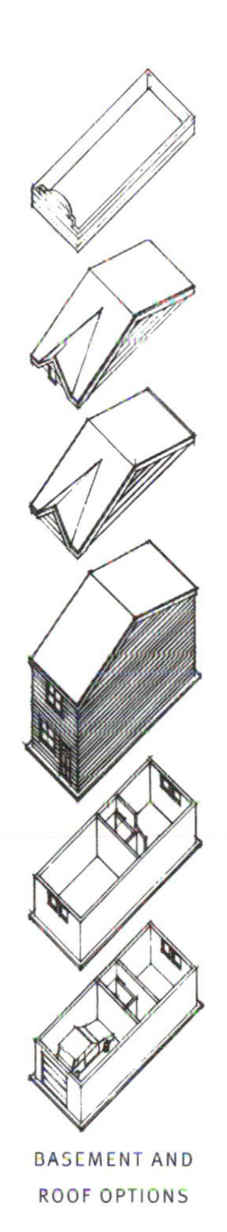

BASEMENT AND ROOF OPTIONS

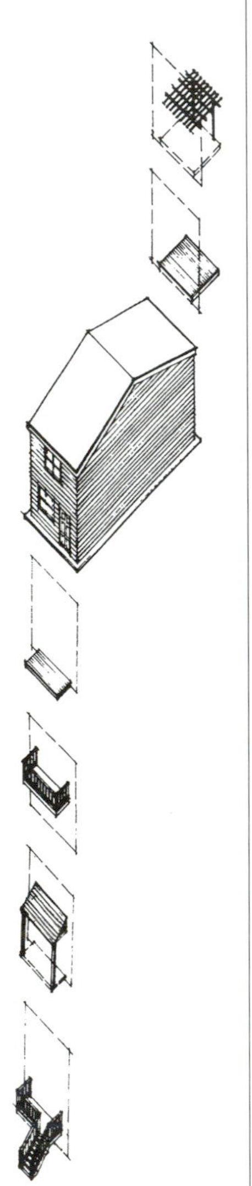
EXTERIOR OPTIONS

Architectural Configuration	$ (1990)
Unfinished basement	4,246
Unfinished basement with garage	5,720
Upper level balcony	346
Lower level balcony (no basement)	300
Balcony steps (with basement)	64
Porch roof	300
Rear deck	720
Rear pergola	419
Sloping roof dormer	872
Sloping roof dormer with window	1,000
Flat roof	2,016
Fully partitioned upper (incl. bathroom)	5,218
Exterior Finishing	
Canexel siding (per facade)	440
Brick (per facade)	522
Stucco (per facade)	522
Vinyl-clad wood double-hung window (unit)	61
Vinyl-clad wood patio door	134
Mullions (per window)	10
Architectural moldings at eaves & door	388
Interior Finishing	
Vinyl tile throughout lower level	282
Vinyl tile in kitchen & bath, carpet in living	330
Carpet on lower level (except bath)	374
Wood-strip floor on lower level (except bath)	641
Vinyl tile on upper floor	825
Carpet on upper level	977
Painted pine stairs	80
Varnished oak stairs	940
L-shaped kitchen in pine	280
U-shaped kitchen in melamine	592
U-shaped kitchen in pine	1,086
Kitchen provided by owner	-774
Pine bathroom vanity	110
Pine bathroom provided by owner	-300

Figure 2 Grow Home options

or partially or fully finished. Exterior finishing for the facade includes vinyl siding, stucco, Canexel (wood) siding, or brick. Windows are wood or vinyl-clad wood. Interior finishing includes a choice of sheet vinyl, vinyl tile, carpeting or wood-strip flooring; different sizes and qualities of kitchen cabinetry, or no kitchen at all; and stairs in carpeted spruce, painted pine, or varnished oak.

We also designed four variations of the base model, which we named after well-known chairs (figure 3). The construction cost of the "plain vanilla" home – the Straight-Back – was $36,912. The second alternative, the Windsor – the prototype built on campus – cost $42,808 to build. This model offered larger and more expensive doors and windows, a balcony and a dormer. We included an unfinished basement in the design of the third alternative, the Ottomane, as well as balconies at the front and rear of the lower floor. The finishes of this model were slightly better than those of the previous two and they brought the cost of the unit to $48,212. The Club was the top-of-the-line model and included a basement with the front part serving as a garage and a room in the rear. The main floor had a kitchen, bathroom and living room, while the upper floor was finished and included two bedrooms and a bathroom. The exterior was clad with brick and had upper and lower balconies in the front and rear, as well as a dormer. The construction cost was $57,225. Since a basement could not be built on campus, we decided to use the Windsor model as an opportunity to display many of our cost-reduction ideas (figure 4). Once we chose it as our

FIRST ALTERNATIVE: THE STRAIGHT BACK

The base model: a plain vanilla, no frills home – the bottom of the line. Construction cost: $36,912

Configuration: Slab-on-grade, no basement; no balconies; upper area unpartitioned.

Exterior: Vinyl siding, both facades; all-wood windows; vinyl patio doors.

Interior: Sheet vinyl flooring on ground level; unpainted plywood floor on upper level; spruce, carpeted stairs; L-shaped, melamine kitchen

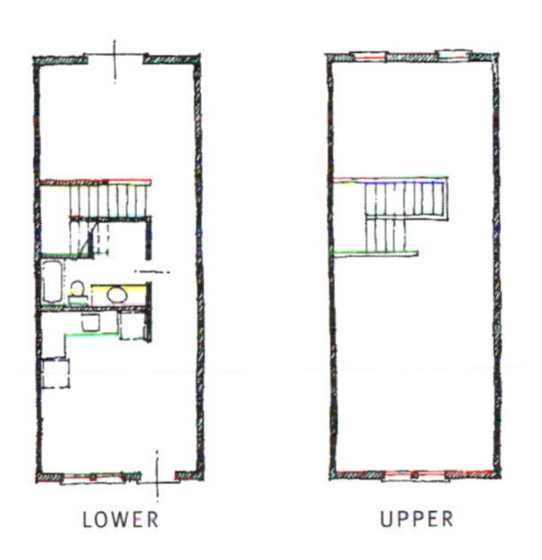

SECOND ALTERNATIVE: THE WINDSOR

This combination was included in the demonstration home as built on the McGill campus. Construction cost: $42,808

Configuration: Slab-on-grade, no basement; rear deck and pergola; lower front porch; upper balcony at front, upper area unpartitioned.

Exterior: Stucco, both facades; vinyl-clad wood windows with mullions; vinyl patio doors; decorative architectural moldings; dormer.

Interior: Vinyl and carpet flooring on ground level; unpainted plywood floor on upper level; spruce, carpeted stairs; no kitchen, provided by owner; no bathroom vanity, provided by owner

Figure 3 Alternative Grow Home designs

THIRD ALTERNATIVE: THE OTTOMANE

A slightly larger model, with basement.

Construction cost: $48,212

Configuration: Unfinished basement; lower balconies front and rear; no balconies at upper level; upper area with single partition; flat roof.

Exterior: Canexel siding, both facades; vinyl-clad wood windows; vinyl-clad patio doors; window shutters.

Interior: Vinyl and carpet flooring on ground level; unpainted plywood floor on upper level; pine, painted stairs; L-shaped, pine kitchen

FOURTH ALTERNATIVE: THE CLUB

A fully finished model with basement garage – the top of the line. Construction cost: $57,225

Configuration: Unfinished basement with garage; upper and lower balconies, front and rear; upper area partitioned; upper bathroom; vestibule.

Exterior: Brick, both facades; vinyl-clad windows; vinyl-clad patio doors; dormer with window.

Interior: White cherry wood strip flooring on ground level; vinyl tile in bathroom; carpeting on second level; varnished oak stairs; U-shaped pine kitchen

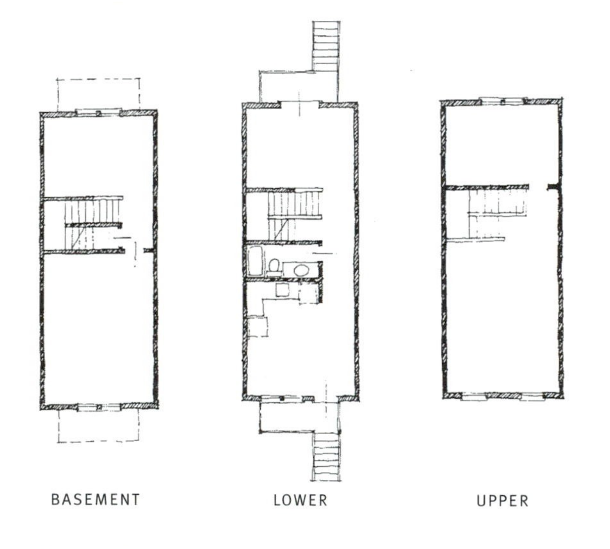

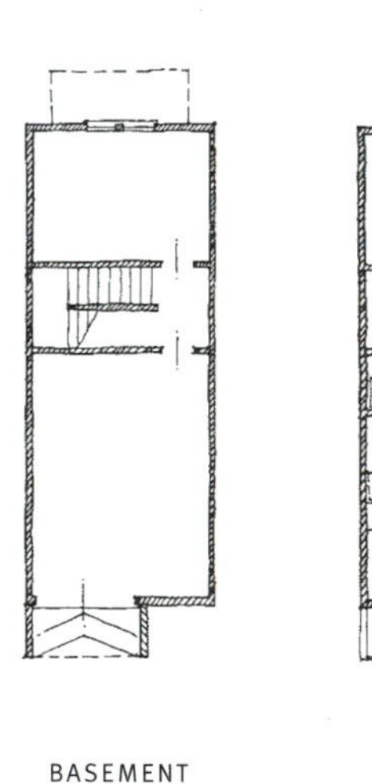

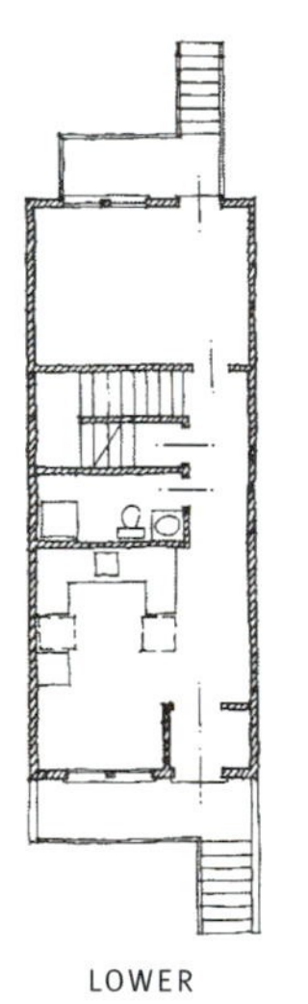

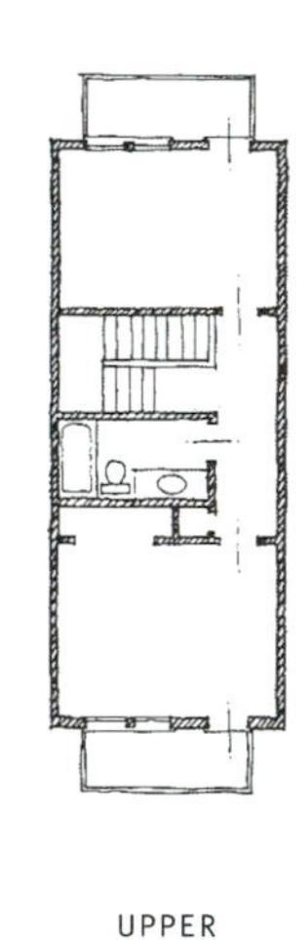

Figure 3 Alternative Grow Home designs

Figure 4 The designers of the Grow Home, Avi Friedman (left) and Witold Rybczynski, on the front balcony of the demo unit

Figure 5 Plans of the demonstration Grow Home

demonstration unit – our prototype – we directed our energies into its construction (figure 5).

We settled on prefabrication rather than conventional, stick-built construction since time was an issue and we could not turn the university campus into a construction site. We chose a panelized approach after visiting different prefabrication plants and examining their building systems (modular and panelized). Aside from the time and price advantage, the panelized building process is not much different from conventional building practice and would enable us to exhibit a construction process that would not be limited to prefabrication. We engaged Modulex, a prefab manufacturer from the Quebec City area. They took about a week to construct the panels in the plant, which they then trucked to the campus to assemble into the rough structure in only two days (figure 6). They built a staircase alongside one of the long walls of the home to let visitors leave the unit from the second floor. Then the work on the interior finishes began, which took ten days to complete. For furnishings, we had to consider that very little in the way of interior decor options was available on the market for a home like ours. Most North American furniture manufacturers and retail outlets subscribe to the cult of Big. But a sense of scale in a home is set by the decor. A bulky sofa creates a small living room. An oversized dining table creates a cramped dining space. We obviously had to look somewhere other than the mainstream furniture outlets. When we visited the Ikea store that had recently opened in Montreal we knew we had found the missing scale. Founded in 1941 in Sweden by Ingvar Kamprad, Ikea stood as much for a philosophy of furnishing as it did for its quality. The do-it-yourself strategy which was a part of that philosophy suited the objectives of the Grow Home. Ikea's catalogue allowed a buyer – at home or in the store – to select a certain model of a type of furniture. A well-illustrated kit of parts for this particular model gave buyers the freedom to select only those parts they needed and could afford. Customers could choose an item in the showroom, take the package of unassembled pieces from the store's warehouse, pay for it, and drive the package home where they would assemble the furniture by themselves, saving assembly and transportation costs.

We produced a detailed imaginary scenario of occupants for the Grow Home to help Ikea's decorators with their task of furnishing the home. The occupants would be a young couple who had rented for a long time and finally decided to purchase a home: a Grow Home. To lower the purchase price, they selected a model with an unpartitioned second floor. A child was born shortly after they moved in. The demo unit as it appeared on campus in June 1990 was predicated on this domestic scenario.

The kitchen was located in the front of the house (figure 7). There were no doors on the kitchen cabinets over the sink; these, it was assumed, could be purchased and installed at a later date by the occupants themselves. Additional cupboards could also be installed on an adjacent wall where a wooden pot rack was placed in the meantime. Similarly, the vinyl floor tiles and the backsplash wall tiles of the kitchen could be replaced with ceramic tiles if desired and when means became available. The dining area was also located in the front of the unit, the table was placed next to a large window which would let in ample light and allow the family to observe the street whenever they ate. The table itself was expandable: when company came over, it could seat eight people. A formal dining room would be rarely used by a family with toddlers, and we reasoned that the absence of such a dining room would be the kind of trade-off that our young couple would be willing to make.

The hallway leading to the rear of the unit was 1.2m (4 feet) wide, a dimension that allowed 30cm (1 foot) to be used as storage space (figure 8). Designing for a compact unit such as the Grow Home meant allocating every bit of space with the same eye for efficiency that one would use in designing a yacht.

Figure 6 Construction of the demonstration Grow Home on McGill campus by Modulex Inc. Clockwise from top: Constructing the foundation platform; assembling the prefabricated wall panels; assembling the second-floor panels.

Clockwise from top

Figure 7 The kitchen was designed for upgrading: doors and additional cabinets could be purchased and installed at a later stage, and the vinyl tiles could be changed to ceramic if desired and when means became available.

Figure 8 Storage cabinets were installed in the hallway to make efficient use of space.

Figure 9 The bathroom was also designed for upgrading: back-to-back plumbing with the kitchen saved $350 in construction cost.

Storage becomes a true challenge in small homes. In northern countries such as Canada, there are two sets of clothes for every household member as well as leisure equipment. Some of this might be stored in the basement, but other goods such as bedding and linens require more regular access. In a small home, a corridor 1.2m (4 feet) wide can even accommodate a small study or home office, or a washer and dryer. In our scenario, the occupants purchased the hallway cupboards after moving in and installed them themselves. The three cupboards were attached to the long common wall that created the corridor, spaced 1.2m (4 feet) apart. Bookshelves made of simple boards were installed in these spaces between the cupboards. The whole wall system added to the acoustic insulation which the wall itself provided.

The bathroom, in the middle of the ground floor, was positioned to share the same plumbing stack as the kitchen as well as any future second-floor bathroom (figure 9). This practice has proved to be a great money-saving strategy, not only letting the plumber use the same main pipes but cutting the installation time. When the sink, toilet, and bathtub share the same wall as the sink and dishwasher in the kitchen, savings are maximized. The bathroom was also designed for future upgrading: ceramic floor and wall tiles could be installed at a later stage. The room included space for a compact washer and dryer stacked to make room for a hot water tank.

The hallway led to the rear of the ground floor where the living room was situated (figure 10). We considered this space to be a guest room as well; the selected sofa was a hide-a-bed. The dimensions of the room – 4.3 by 3.7m (14 by 12 feet) – made for a

Figure 10 Large French doors let in light to the living room and provided access to the back yard.

comfortable space that allowed room for a sofa chair and a television cabinet. Double French doors let light flood into the room and provided the illusion of greater space. These doors opened onto a rear deck with a pergola which had been selected by the occupants from the catalogue presented to them in the model unit. In the summertime, the deck served as a sitting space, expanding the living room. The carpet in the living room was designed for future upgrading and could eventually be replaced by a wood floor.

Stairs led to the second floor where the space was finished (painted, with carpeted floor) but unpartitioned (figure 11). There was a child's area in the rear, a study in the middle and a double bed in the front. This floor could be used initially as a loft bedroom or as a large family room if the downstairs area in the rear was used as a bedroom. With the passage

Figure 11 The family in the demonstration unit scenario was a young couple with a baby. They planned to convert the unpartitioned second floor into two bedrooms and a bathroom.

of time, once the second storey was completed, this upper space could provide a master bedroom, a child's room, and a second bathroom. The room could easily be partitioned in the future by extending the half-wall by the stairs into a full wall and including a door to create front and back rooms with a bathroom located centrally (figure 12). As a cost-saving measure, no clothes closets were built (as is similar to the practice in European homes) since chests of drawers would be used for that purpose.

The construction time did not allow us to use brick on the facade. Brick is a common cladding material in both old and new Montreal housing, reflecting a sense of permanence highly valued by home buyers. Instead we chose stucco, a durable and low-maintenance material with an attractive appearance. There was no good reason, we figured, that a small affordable home should have to look mean or cheap. First-time buyers made their choice of a home with a great deal of prudence and would be unlikely to gamble on a new or untested product or technique. They often associate their new home with the home they grew up in and seek a traditional appearance. For ornamentation on the Grow Home prototype, we employed a variety of classic design details including a Georgian-style strip of molding over the front door, a dormer roof and mullioned windows (figure 13). Our occupants selected that dormer along with an upper balcony, an extension of the second-floor loft. Balconies form part of the vernacular architectural language in the city of Montreal, a city with a short summer and a hunger for sun. The campus unit was meant to display one slice of a row of similar homes. Different choices of facade components selected by different occupants would bring about a lively streetscape. The result would not be the usual monotonous row associated with lower-cost housing projects.

The campus demonstration unit struck a chord and the Grow Home received an avalanche of media attention (figure 14). Local, national, and international media outlets were eager to tell the public about the new design. The *New York Times*, *Los Angeles Times* and *Toronto Globe and Mail* carried features on the Grow Home. *Good Morning America* broadcast live from the campus to its viewers. The phone calls and requests for information did not subside for quite some time. We had underestimated the public's desire for affordable housing solutions. As a result, approximately ten thousand people visited the demonstration unit. At busy times, such as during the lunch

Figure 12 A model showing space arrangements of the second floor – before (left side) and after (right side)

Figure 13 Elevations of the demonstration Grow Home

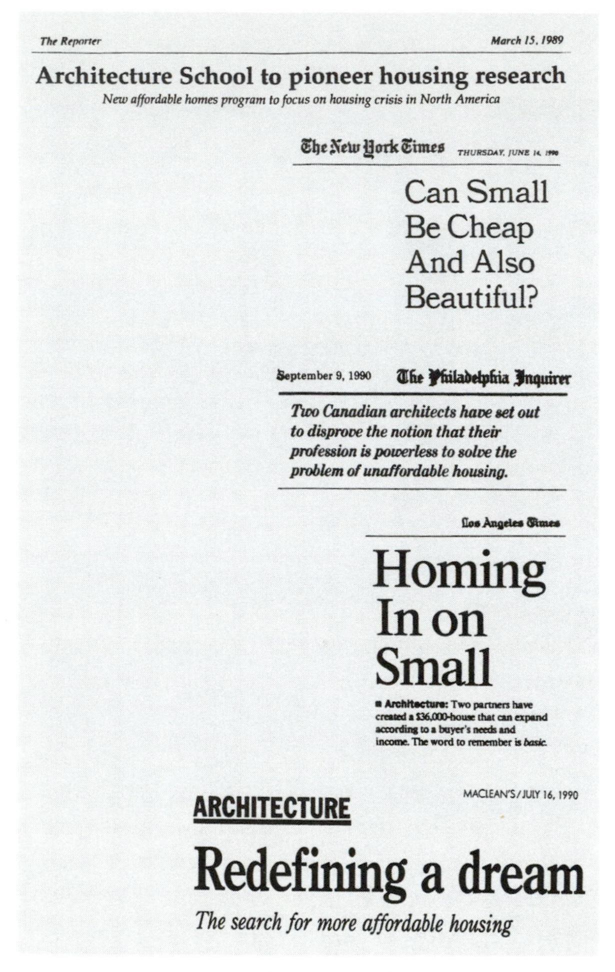

The Reporter — *March 15, 1989*

Architecture School to pioneer housing research

New affordable homes program to focus on housing crisis in North America

The New York Times — THURSDAY, JUNE 14, 1990

Can Small Be Cheap And Also Beautiful?

September 9, 1990 — The Philadelphia Inquirer

Two Canadian architects have set out to disprove the notion that their profession is powerless to solve the problem of unaffordable housing.

Los Angeles Times

Homing In on Small

■ **Architecture:** Two partners have created a $36,000-house that can expand according to a buyer's needs and income. The word to remember is *basic.*

MACLEAN'S / JULY 16, 1990

ARCHITECTURE

Redefining a dream

The search for more affordable housing

Figure 14 Sample of media coverage of Grow Home demo unit

hour, they lined up in front of the home, patiently waiting to see what all the fuss was about.

We asked 636 of these visitors, selected at random, to complete a questionnaire designed to elicit their views on the home's main features. The questionnaire was structured around the theme of the trade-offs that potential buyers would be willing to make in order to become homeowners. We were particularly interested in the relation between income, the presence of children, and the home-buying decisions of potential purchasers. The questionnaire was brief, with a total of eight questions on the visitors' opinions of the traditional appearance of the demonstration unit, the unfinished second floor, the compact size, the projected price, its viability as a suburban home, and the quality of the interior finishes, as well as pertinent facts such as age, income, and presence of children.

The demonstration unit attracted a wide diversity of visitors. We divided those who filled out questionnaires into two groups, based on age. The first and largest group was made up of young (under the age of thirty-five) first-time home buyers who were actively looking to buy a house or planning to do so in the near future (75.5 per cent). The second group included all those who already owned a home: people between the ages of forty-five and sixty-four (20.0 per cent), and retirees who planned to trade a large home for a smaller one and who were interested in a housing unit like the Grow Home (4.5 per cent). The visitors could also be divided into two groups based on income level: those who earned under $40,000 (just over half) and those who earned over $40,000 (just under half) (table 1). Almost two-thirds of the group who earned under $40,000 indicated that they were actively looking to buy their first home.

In an earlier survey of home builders in the Montreal area, I had found that when first-time home buyers begin to look for a house they generally search for one beyond their means (Friedman 1987). Such buy-

Income (1990 $)	Profile of respondents by income (%)	Willingness to live in a unit which is smaller than 100m² (1,000 ft²) (%)		Willingness to drive 30 minutes beyond current homes (%)		Perception of quality of finishes (%)			
	Respondents	Willing	Unwilling	Willing	Unwilling	Excellent	Good	Acceptable	Disappointing
20,000 – 30,000	29.1	86.3	13.7	68.8	31.2	17.9	46.7	32.6	2.8
30,000 – 40,000	23.9	81.7	18.3	69.2	30.8	19.5	46.0	34.3	0.2
40,000 – 50,000	18.1	73.1	26.9	71.1	28.9	18.7	42.7	36.4	2.2
over 50,000	28.9	60.0	40.0	66.6	33.4	17.3	49.3	31.4	2.0

Table 1 Relation between income level and respondent profile, willingness to live in a smaller unit, locational preference, and perception of finishes' quality

ers scale back their aspirations once they realize that they cannot afford what they want. We wanted to know whether potential home buyers would be willing to live in a home smaller than 100m² (1,000 ft²), such as the Grow Home, given its low suggested price. We found that roughly 84 per cent of respondents with incomes under $40,000 were willing to trade space for affordability. This percentage drops to approximately 60 per cent for respondents with incomes over $40,000 (table 1). The willingness to accept a smaller space was due, we reasoned, to the buyers' realization that the chances of owning a home in the city would be slim given the home prices at the time. It could have also indicated a trend towards smaller city dwelling units due to socioeconomic changes and the reduction in household size. Potential buyers preferred a single-family detached home as their first choice, but they would accept a small townhouse as a compromise since it allowed them to live in the city and reduce their commuting time.

Other researchers who have studied the relation between home-buying decisions and locational preference have identified the importance of proximity to service facilities (i.e., education, health, shopping) among households with children (Stapleton 1980). We also decided to find out whether buyers would be willing to trade location for affordability. The visitors to the Grow Home demonstration unit were asked if they would be willing to travel an additional thirty minutes to their places of work and to service facilities in exchange for buying a unit with a price of $75,000 (by comparison, the average new home in Montreal at the time was $120,000). We found that over two-thirds of all respondents would be willing to trade locational preference for affordability (table 1), which showed that active buyers realized that only a limited number of home-buying options were available to them. Trading away location was one option that many were willing to accept.

We also wanted to know whether households

Respondents age 18–45	% Satisfied	% Dissatisfied
Without children	61.7	38.3
With children	27.4	72.6

Table 2 Relation between presence of children and satisfaction with small unpartitioned space

with children would accept the cost reduction strategies associated with the Grow Home. Previous studies had demonstrated that concern over the number of rooms and amenities is especially important in the decision-making process of households with children (Kobrin 1976). We therefore investigated the relation between the presence of children and the willingness to buy a 100m^2 (1,000 ft^2) unit with an unpartitioned second floor. The respondents were made aware of the fact that the second floor could eventually be subdivided to contain two bedrooms and an additional bathroom. Unsurprisingly, almost three-quarters of the respondents with children indicated that they would not be satisfied with a small unit that included an unpartitioned second floor (table 2). Novel ideas that do not fit the needs of the group for which they are intended have little chance of acceptance in the marketplace. Furthermore, the provision of privacy for household members is a minimum requirement for many first-time buyers when they are given a choice between renting and owning.

Another concept fundamental to the Grow Home was the design for future upgrading of surface finishes. The original design recommended that the initial materials be inexpensive yet of reasonable quality. The walls in the kitchen and bathroom, for example, were painted; the assumption was that buyers could tile them in the future. We wanted to discover how the suggested finishes would be perceived by visitors from different income groups. No discrepancies were found between the various income levels (table 1), leading us to conclude that most people – no matter what their income level – expect lower-quality finishes in a low-cost house. Home buyers associate price with quality and they are ready to accept lower quality for lower price. We were encouraged to find, though, that over nine-tenths of the respondents liked the idea of completing the second floor themselves and upgrading the finishes.

The results of the questionnaire confirmed our initial assumption that first-time home buyers were ready to make significant trade-offs in order to become owners. Once potential buyers faced the reality of the housing market, they would necessarily *and* willingly leave behind some of their original, unrealistic expectations.

We had hoped that – in addition to the media and the general public – the Grow Home would attract the attention of builders. They are, we knew, the key decision makers with respect to the acceptance of new ideas. Of course, a narrow townhouse is not new in architecture: many of them have been designed and built in Europe and North America over the centuries. It was the ingredients that we brought to the design process, as well as the timing and context of their presentation, that made our approach novel. We were eager to find out what the industry verdict would be.

4 FROM CAMPUS TO SITES

Public reaction to the Grow Home demonstration unit was a surprise. The extensive media coverage was an indication of great interest in the subject. We had underestimated society's desire to keep the dream of home-ownership within the grasp of many. The coverage generated almost 1,300 written requests for information from across North America. About 20 per cent of the requests came from developers, builders, real estate agencies, and other construction-related organizations. We knew that it was this group who would determine whether the Grow Home would find its way onto a building site or remain a prototype.

The Montreal home-building industry was in the grip of a deep recession when builder Leo Marcotte walked into my office in January 1991 (figure 1). Interest rates were at around 13 per cent and activity had drawn to a standstill. Marcotte owned a small building firm at the eastern tip of the island of Montreal, an area called Pointe-aux-Trembles. The district was a mix of single-family homes of the type that Marcotte himself built and four-to-six-storey apartment buildings which rented mostly to young couples with or without children and to a lesser extent to singles and single-parent families. Montreal is known in Canada as a city with a high percentage of renters, as a place where it is convenient to rent. Stable demand, steady rents, and a large inventory as well as a wide range of housing stock have made it a haven for tenants. Those renters who wished to buy, however, could not save the necessary cash for a down payment on a new home, and renting became a way of life for many. All that changed in the mid-1970s. Paradoxically, a key factor was the establishment of a rental board, a government agency that oversaw the rental market. Strict regulations limited drastic increases of rent. Builders of rental properties soon realized that there was no economic advantage in building this particular kind of accommodation. Rents on built units went up. A two-bedroom unit,

Figure 1 Leo Marcotte, builder of the first Grow Home development

which most young couples desired, went for between $500 and $700 per month, depending on location.

Marcotte had come to McGill to take part in one of our studio projects. When he saw the Grow Home design, he was sceptical about the possibility of constructing a marketable home in Montreal for $46,000. I understood his concern. The home-building industry is known to be conservative and highly reluctant to embrace innovation. A home builder who was once asked if he was interested in an idea that would reduce construction time answered that he would be ready to implement any new idea with twenty years of practical experience. Concepts, techniques, or products that change traditional routines are carefully scrutinized. What are the reasons for this reluctance? Why has the building industry not followed patterns of innovation set by the computer, electronics, or bioengineering fields?

It is perhaps an exaggeration to claim that no new ideas find their way into home-building practice. In the past half-century, several products and techniques have changed the way homes are put together. Prefabricated roof trusses, gypsum wallboard, plywood sheeting, and batt insulation were revolutionary items at the time of their introduction and are commonly used by builders today. It took a long time, however, for these products to be widely accepted. A researcher with the National Association of Home Builders Research Foundation commented that it takes between three years ("overnight") and fifteen years ("a mean time period") from the time a technique or product is introduced until it gains widespread acceptance in the industry (Sichelman 1987). In order to understand acceptance or rejection of innovative ideas, one needs to be familiar with the working of a typical home-building firm.

Due to fluctuations in housing demand – a result of economic trends – most housing developers have small and often family-controlled firms. Over 80 per cent of these firms have fewer than five full-time employees. In a highly competitive industry where profit margins range from between 10 and 15 per cent, developers need to keep their overhead expenses to a minimum. Capital investment in land holdings or a large staff are not options if the firm is to stay in business. In their effort to reduce costs, developers assume many roles, acting as the composers and conductors of a large orchestra. Land acquisition, financing, product development, obtaining permits, marketing and sales, and of course supervising construction are some of their many personal responsibilities. The typical development firm has become removed from the actual building process itself. The construction work is subcontracted to highly specialized subtrades. Once underway, the construction process itself is uninterrupted. Trade after trade will arrive at the building site to do their parts. Developers much prefer designs that use standard products in an accustomed procedure. A design that introduces products unfamiliar to the trades and specifies unconventional uses of them will bear higher costs. A project that includes a repetitive set of tasks will be welcomed. Mass construction will lower costs. Purchasing large quantities of the same products will gain a discount from a supplier of materials. The installation of these standard products with minimum variations by the trades will result in a shorter construction period and lower costs. Familiarity with products and methods of installation have therefore become essential to the development of design strategies for cost reduction. Imposing a new technique on a subtrade or changing a routine that does not contribute to the efficiency of the entire process will therefore not be looked upon favourably.

Financial institutions also hold great power over the builder. Decisions concerning the flow of funds into the home-building sector are made strictly on business terms. A banker will be reluctant to lend money to a small development firm that wants to experiment with a new technique or products. Such firms lack the necessary means to correct an expensive large-scale failure if one occurs.

On the buyer side of the equation, first-time home buyers are also typically reluctant to assume risk. They are likely to buy a home from a reputable home builder with a good track record. They are less likely to make the largest investment of their lives in an untried design concept with an unconventional roof design, for example, that could end up leaking.

So what makes builders change their minds and buy into a new concept, technique or product? A clear demonstration of cost reduction would be an incentive, offering a financial edge over competitors in a tight market. A significant time saving would also be an incentive, primarily in a boom period when demand exceeds supply. At the opposite end, stagnation and reduced demand might also tend to stimulate new thinking and risk taking. Builders will do practically anything within their power to draw hesitant consumers to the market.

Marcotte therefore decided to test our claim and to weigh his risk. Despite the complexity of a built house, making an estimate for one is fairly simple. A brief examination of the plans reveals how big the house is, the number of bathrooms, the quality of finishes, the presence of an indoor garage, and how complicated it will be to put it all together. These indicators lead to a per-square-foot cost. When Marcotte totalled his numbers he knew that we were in the right range (table 1). When he added the cost of land to the construction cost, he was able to calculate a monthly mortgage payment: approximately $500, the amount that renters in the area were already paying. That side of the deal was clinched, and it turned out to be an easy sell. But Marcotte was not sure about the design itself. Montrealers are familiar with row-

Item	Unit	Typical Home		Grow Home	
			Cost ($)		Cost ($)
Land					
Lot size	Feet	50 x 100		14 x 100	5,600
Lot area	Sq. ft	5,000		1,400	
Cost of land (unserviced)	$/Sq.ft	4		4	
Total cost	$	5,000 x 4	20,000	1,400 x 4	5,600
Services					
Lot frontage	Linear ft	50		14	
Cost of services	$/Linear ft	450		450	
Total cost	$	450 x 4	22,500	450 x 14	6,300
Construction					
Unit dimensions	Feet	30 x 40		14 x 36	
Area per floor	Sq. ft	1,200		504	
Number of floors		3		3	
Area of finished space	Sq. ft	3,600 (1,200 x 2)		1,800 (504 x 2)	
Cost of finished space	$/Sq.ft	50		37[1]	
Cost of construction	$	50 x 3,600		37 x 1,008	37,296
				18.5[2] x 504	9,324
Total construction cost			180,000		46,620
Marketing					
5% (of construction)	$	5% x 180,000	9,000	5% x 46,620	2,331
Sub-Total			231,5000		60,851
Overhead and Profit					
15% (of construction, marketing and serviced land)	$	15% x 231,500	34,725	15% x 60,851	9,127
Total	**$**		**266,225**		**69,978**

1 Based on a survey of seven projects. The cost was reduced due to simplification of unit configuration, elimination of bearing walls, back-to-back plumbing, same-size floor joists and roof trusses, common walls, only two facades.

2 Construction cost of unfinished space is half of finished space.

Table 1 Cost effectiveness of the Grow Home (1990 $). Both the Grow Home and the typical metropolitan Montreal home are assumed to be constructed in the same location with similar construction methods and materials.

housing, but who knew if first-time buyers would be willing to buy a unit only 4.3m (14 feet) wide? Before committing himself to building a Grow Home, Marcotte wanted to have a closer look at the design.

Marcotte picked me up on a snowy day in February 1991 at the Honoré Beaugrand subway station in the east end. We drove in his Nissan Pathfinder to a strip mall where he suggested we have lunch. "I have these sixty-by-hundred-foot lots," he told me as he unfolded a site plan. "I can subdivide them to fit your Grow Home design and build them in groups of three, four or even eight. I don't have any problem with the narrow width, but knowing the clients in Pointe-aux-Trembles and what they're looking for, I don't think the campus model, the Windsor, would be ideal. Let's look at that catalogue you guys put together." He opened the report he had picked up in our office. It contained four alternative layouts, including several exterior features and their cost. We knew that interior layout, facade design, and facade materials are subject to the preferences of builders and clients that can change from region to region. We had intended the report to become a catalogue of options, much like a do-it-yourself furniture sales catalogue. "I think this one would work better," Marcotte said, pointing to our Le Club model. I suggested that an indoor garage would make the unit too expensive, and flipping through the pages, asked him to consider outdoor parking and a habitable basement. "There's an idea!" Marcotte exclaimed. "Let's use the second floor of Le Club and add an unfinished basement. Then you've got your grow space." On the back of a place mat he made a series of quick calculations. Finishing the second floor would cost him $5,000 and he could still keep within his desired price range.

Marcotte knew his market. He had met many of his potential clientele in his large model unit and talked with them about their jobs and income, their family plans and even their hesitations as he walked them through the model. He also knew what they would be looking for in a home: what would be a priority space and what could wait for later, which of the interior functions would be a must and which could be a compromise. "We also need to do something about the main floor," he said. "A kitchen at the front of the unit may work in other communities but not in Pointe-aux-Trembles."

He also drew the line on the amount of innovation that he would be willing to introduce and the degree of risk he would undertake. "No problem," I said. "We used a short span to allow for flexibility. Let's just flip our original layout: the living room can move to the front and the kitchen can go to the back." I realized that there was merit in the switch. Placing the kitchen in the rear would let parents working in the kitchen watch the kids playing in the backyard. Eating outside on the back porch would be made much easier. There would be no need to carry a coffee mug and plate and a newspaper under one arm across the whole house. The proximity made sense.

We passed the pen back and forth and continued to add lines on the crowded paper placemat. Once the kitchen was moved to the back, there was no point in keeping the adjacent bathroom. The second floor, offered as a finished space, would have a complete bathroom, so all we needed on the main floor was a powder room (toilet and sink). It also made no sense to keep the stairs where they were located in the Windsor model, so we hunted for the proper place. If the kitchen moved to the back, the living room would be in the front, the dining room would be adjacent to the kitchen; in order to accentuate the open space, the stairs could be placed along the longitudinal common wall.

It had all come together. "I think it's going to work," said Marcotte. He grabbed our Grow Home publication and opened it. "I like the Windsor facade," he said, pointing to the page with our drawing. "It looks elegant. I think I can clad it front and back with brick and still keep it affordable." He felt that this small investment would make the row look more upscale and eliminate the low-cost stigma. "The end units can also have windows on the long wall," he added. "I might even consider offering a fireplace and nice bathroom fixtures as options."

I reflected on the architectural experience I had just been through. There was no drafting board, computer screen or coloured markers – just the back of a placemat on a greasy table in a tiny restaurant. I compared this experience with the study of architecture, my practice, and my teaching at the university. Had I compromised my professional integrity? Should I have been more stubborn and resisted change? I don't think so. Marcotte had to sell a product to people he knew, and he had to be comfortable with what he sold. Had the principal characteristics of the design changed? No. It was still a small, narrow-front townhouse with an indoor space for expansion, a well-built home at an affordable price.

I still knew that many architects would be reluctant to participate in this type of working relationship. Few architects do, in fact. Many "serious firms" do not participate in the design of mass-produced developments because of their inability to maintain sufficient control. Their original plans are frequently changed without consultation, and payment – especially in difficult times – may be delayed by builders. This situation has brought about the creation of specialized house designers, primarily in the single-family housing sector, who operate outside the formal professional structure of architecture and who are often not registered architects. They charge far below recommended rates for a single plan that is used numerous times in different projects (Blau 1984; Montgomery 1977). But the design of housing is a missed opportunity for architects in an industry that builds approximately two million units in the United States and some 150,000 in Canada per year. (The type of services they are asked to provide in addition to design may even include the graphics on promotional material and the decor of a model unit.)

Big challenges still lay ahead for Marcotte. He would have to take the plunge: subdivide his land, get construction drawings ready, talk to his banker, and advertise the project. It was a lot to do in a short time. In Montreal, people buy homes in the winter for July occupancy. Winter conditions mean that construction can begin only in late March. Marcotte had to find a way to get his clients' attention. He had a bright idea: he would emphasize the monthly payments which would include mortgage, municipal taxes and heating. Because the monthly amount would be equal to or only slightly higher than what they were already paying in rent, interested potential buyers would at least visit to satisfy their curiosity. Marcotte rushed to his printer and ordered flyers with the message "*Incroyable! $479/mois!*" He had them delivered to all the rental apartment buildings in Pointe-aux-Tremble, and he also placed an advertisement that weekend in the home section of a local newspaper which prominently featured the monthly payment (figure 2). The outcome was still unknown. Would buyers notice the ad? Would the dollar figures be convincing? Would they be able to afford it? Would they come to the sales office?

And so it was that on a snowy Sunday morning in February 1991 I drove out to a project site. About fifteen cars were parked in a disorderly fashion around a temporary office on the site. Inside the office, couples in their late twenties and early thirties, some with toddlers, were lining up. Floor plans based on the original design concept were stuck to the office walls. "It's unbelievable," Leo Marcotte whispered when we met. "I've never seen anything like this in my life." He sold twenty-four units that weekend. The entire development of eighty-seven units was sold in two weeks. He quickly started to sell phase two with 250 additional Grow Home units. His success was also due, in part, to an incentive program that was put in place to assist would-be home buyers and to rescue a stagnant industry from recession. The provincial and municipal governments implemented programs to reduce the cost of home-ownership for the first few years, while the federal government made the purchase possible through an existing mortgage insurance system that allowed a down payment as low as 10 per cent of the home's entire price. In an effort to promote housing starts, the provincial government offered financial aid to home buyers purchasing a new dwelling built by local builders with a price of $150,000 or less. Applicants had the option of choosing 8.5 per cent financing for three years or a subsidy of 4.5 per cent of the total cost of the unit up to a maximum of $5,000. With mortgage rates at the time averaging 12.5 per cent, the program represented monthly savings of up to $208 for an $85,000 unit. The City of Montreal implemented its own program with the objective of attracting residents back into the city. An annual reduction in property tax of $1,000 a year for five years was made available to first-time buyers purchasing a newly built or renovated unit costing no more than $100,000 (not including land, infrastructure, or sales tax). The city also offered automatic acceptance to the program if the applicant qualified for the provincial subsidy. The credit helped reduce the monthly carrying cost for starter homes by an additional $83. These programs and incentives enabled a buyer to purchase a $76,000 home for $618 per month (including sales tax).

Figure 2 Media advertisement for the Grow Home, emphasizing the low monthly payments.

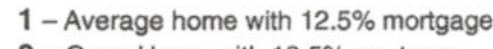

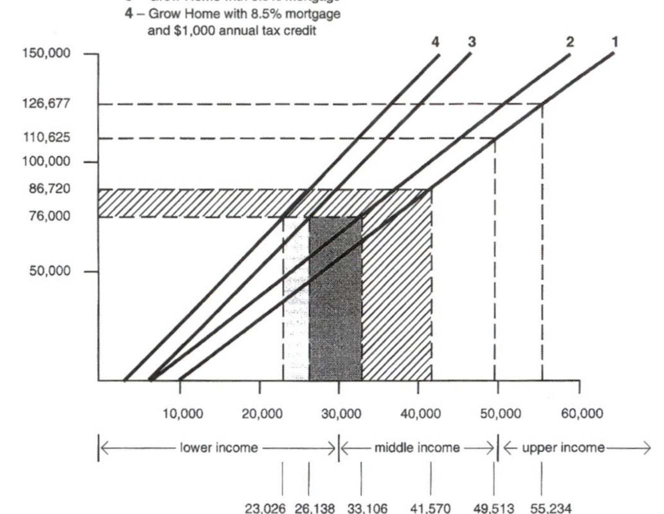

* Based on total carrying costs, including capital, interest, property tax and heating. Annual property taxes and heating costs for the Grow Home were estimated at $1,115 and $240, respectively. Costs for other types of units were based on averages for the City of Montreal and assumed to be $2,115 for property taxes and $980 for heating. Sales taxes were not included. A gross debt service ratio of 32% is assumed with a 10% down-payment.

Figure 3 Combined effect of government subsidies on housing affordability (1990 $)

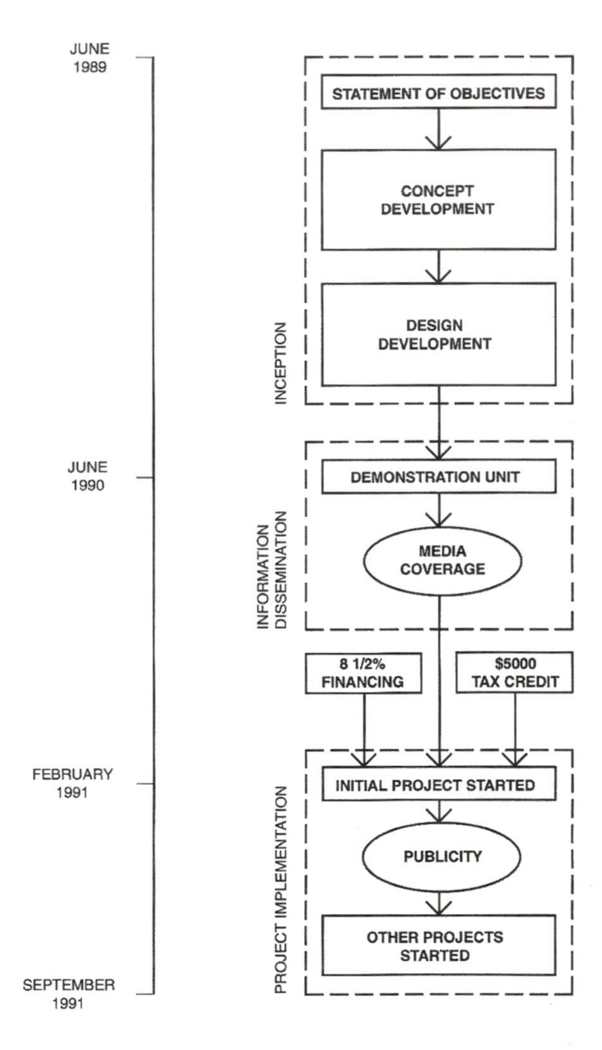

Figure 4 Sequence of events in the Grow Home project

The average household income in Montreal at the time was approximately $41,500. Assuming a 32 per cent gross debt-service ratio and a 10 per cent down payment, the maximum affordable price for a home at that income would be close to $86,000. This price was $24,000 lower than the average price of a resale listing and $40,000 less than the average price of a new home. At those prices, household incomes of $49,500 and $55,300 respectively would be required to make a purchase – incomes above those of the target markets for Grow Homes. Through lower property taxes and heating costs, the Grow Home shifted the affordability curve (towards the left in figure 3, line 2). Because of the lower development and construction costs, the units were priced at $76,000, placing them within the range of households earning slightly more than $33,000 (just below the middle income bracket). With the offer of 8.5 per cent financing, the provincial government put the units within the range of households earning $26,000. The tax credit offered by the city shifted the curve even further (figure 3, line 4). With a down payment of less than $8,000, apartment tenants with a gross annual household income of $23,000 (below the poverty line in Canada) were able to become homeowners.

The news of Marcotte's success in selling affordable homes in east-end Montreal spread quickly (figure 4). An article in a local paper detailed the process and the outcome. In recession-stricken Montreal, it did not take long for the competition to come by, visit, and begin to copy – typical of an industry with a small investment in market research and the development of new products. Small firms simply do not have the required capital to devote to R&D. The research is often limited to visiting a local home show, reviewing trade magazines, and visiting competitors' sites. Builders also rely on research made by provincial and federal housing authorities and by their own trade associations. Even so, they are still wary of the outcome and they like to let someone else try out an innovative idea. Rogers (1983) suggests that only one in forty builders can be considered an innovator (figure 5). Fewer than one in seven will be quick to follow and implement an innovation. The rest will follow slowly, having ensured that a track record has been established.

Those early adopters made their decision to proceed with building Grow Home projects within two weeks of being exposed to the concept. Most of these builders completed a model unit within two

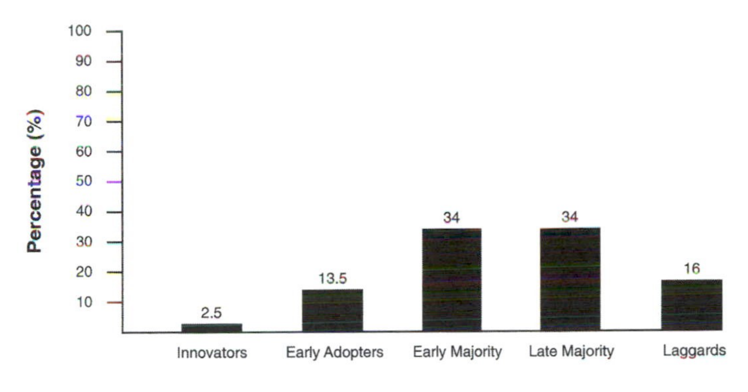

Figure 5 The adopters of innovation (Rogers 1983)

months of their decision to build. This quick action led to some widespread competition in a short period of time. In most of the initial Grow Home projects, the progress of sales was very encouraging. Six of the builders began second projects after their first projects had sold out (figures 6 and 7). Ironically, the bulk of the sales was achieved during a ten-month period of economic recession. Although conservative spending and the lack of investment dollars did not offer any hope of immediate recovery, the situation nevertheless proved to be an asset for several reasons. Builder production had been at a standstill for the preceding fourteen months; this state of affairs encouraged builders to investigate new markets and made them more receptive to innovative design alternatives, particularly when the initial investment in land and building was only a fraction of what they might have otherwise attempted. Most of the builders already owned the land, while others tried to liquidate stock that had been sitting idle for a year and a

Figure 6 The locations of the first Grow Home developments were in small suburban towns where inexpensive serviced land made the design highly affordable.

half. With the increased competition in a struggling labour market, many of the subcontractors also offered reduced rates.

Some of the cost-saving strategies included in the design of the Grow Home were instrumental in attracting the interest of the local builders. The flexibility of the design not only simplified the task of modifying the layout but also enabled builders to offer a wide range of options to buyers without significantly complicating the construction process (table 2). The offer of selecting options was considered to be a key selling point, since it allowed buyers to personalize their homes and adjust the design to suit their budgets. Last-minute changes could be made to the floor plan in response to particular demands, and by keeping the size and location of openings constant, most builders could proceed with the construction of the structure and envelope before finalizing the interior layout (table 3). The 4.3m (14 foot) dimension also represents the cut-off point for a floor structure consisting of 38 by 235mm (2 by 10 inches) joists at a spacing of 400mm (16 inches) on centre. To add 305mm (one foot) to the width would require upgrading to a structure costing 25 per cent more (table 4).

Eleven of the builders followed Marcotte's lead and provided brick veneer on the exterior to increase quality and project an image of permanence, while the remainder used a cement-based aggregate finish. With a marginal increase in the required cost of excavating to a minimum of four feet below grade to ensure that the foundations cleared the frost line, all units were built with unfinished basements which became the add-in space, increasing the overall floor area by $50m^2$ ($500 ft^2$). Indoor garages were included

Builders	**A**	**B**	**C**	**D**	**E**	**F**	**G**
Units built	4	48	93	22	62	177	47
Exterior finishes	•						
Floor finishes	•	•	•	•	•	•	•
Paint colours	•	•		•			
Kitchen cabinets	•	•			•	•	•
Bathroom fixtures	•	•	•	•	•	•	•
Plumbing fixtures	•	•				•	
Lighting fixtures	•	•	•			•	•
Wood finishes	•	•				•	•
Interior railings						•	

Table 2 Options offered to buyers by Grow Home builders

Poured Concrete Foundations:
254mm foundation walls
750mm x 300mm strip footings
100mm slab in basement

Wood-Frame Construction:
50mm x 152mm stud walls
50mm x 254mm floor joists
Prefabricated roof trusses

Common Walls:
Double stud wall; 50mm x 100mm with 2 layers of batt insulation; 4 layers of 12.7mm gypsum wallboard

Insulation Values:
3.8 RSI for the walls
7.3 RSI for the roof

Interior Finishes:
12.7mm wallboard, walls and ceiling
Parquet or carpet flooring
Cushion floor in ground-floor bathroom and kitchen
Ceramic tiles in upper-floor bathroom

Exterior Finishes:
Brick veneer on entire front
Brick and/or aluminum around sides and back

Table 3 General construction characteristics of selected Grow Home projects

Figure 7 Grow Home developments in the Montreal (Quebec) area. Clockwise from top: Notre-Dame-de-Grace; Pointe-aux-Trembles; Longueuil; and Ste Dorothée

in 15 per cent of the homes. Vestibules and walk-in closets were added to the units in one of the projects, while separate garages were added to the sides in another. While the basic dimensions, appearance, and quality of the interior finishes were retained as they were in the demonstration unit, these other changes to the ground-floor plan and the exterior finishes were made to increase the marketability of the home (figure 8).

Subdivision of land into 4.3m (14 feet) parcels was not permitted by any of the municipalities where Grow Home-type projects were built. As a result, the buyers were not able to obtain private ownership of their land. The builders who attempted to obtain the required changes were refused. The majority did not even submit any proposals because they knew that the amount of time it would take to process the application would jeopardize their competitive position. Despite these regulatory obstacles, the builders found a way of putting up the units while remaining within the legal restrictions. The most common solution was to build the units in manageable groups of three, four, or six and sell them under a co-op or condominium arrangement. Even though most of the builders felt that their sales would have been better had they been permitted to sell the homes under freehold ownership, selling the townhouses as condominiums actually helped reduce construction cost by doing away with some of the building code requirements. Under freehold ownership, for instance, some municipalities require that the units be separated by a four-hour masonry firewall. For condominiums, on the other hand, a continuous two-hour fire separation is sufficient. Under such an arrangement,

Construction category	Average cost per unit ($)[1]	Percentage of total cost	Average cost per square metre ($)
Site	2,082	4.5	17.10
Foundations	2,920	6.3	23.98
Common wall	2,228	4.8	18.28
Rough carpentry[2]	11,666	25.2	95.81
Doors & windows	2,536	5.5	20.86
Services	5,699	12.3	46.77
Interior finishes	9,665	20.9	79.35
Furnishings	2,245	4.9	18.39
Exterior finishes	4,689	10.1	38.49
Landscaping & paving	2,563	5.5	21.08
Total	**$46,293**	**100.0**	**$369.89**

1 Material and labour costs in 1990 dollars (does not include land, infrastructure, overhead, and profit). Based on gross floor area including a half basement.

2 Includes thermal and moisture protection.

Table 4 Elemental breakdown of average Grow Home construction costs (1990 $)

builders could use wood and drywall for the common wall instead of concrete block, resulting in savings of about $3,500 per unit. Some municipalities require separate water cuts in the infrastructure when the units are on separate cuts of land, a cost which is absorbed by the builder and passed on to the home buyer. At $3,000 per water cut, the builders could save $3,000 per each group of four units. Similar restrictions in the areas of plumbing requirements and acoustic insulation led to additional savings in some of the developments.

Since most of the projects were erected in new developments, the "not in my backyard" syndrome was not a major factor for any of the builders. Although

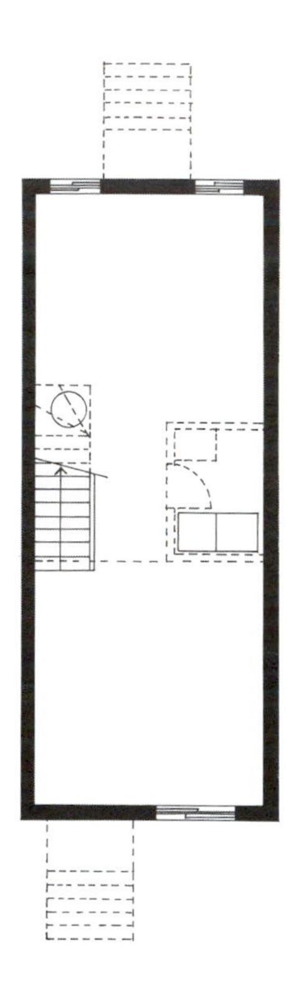

Figure 8 Examples of Grow Home floor plans, as modified by builders

Unfinished basement

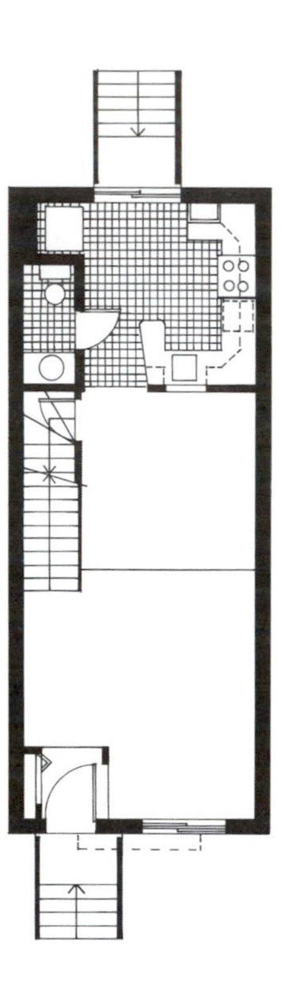

Ground floor with kitchen and powder room in the rear

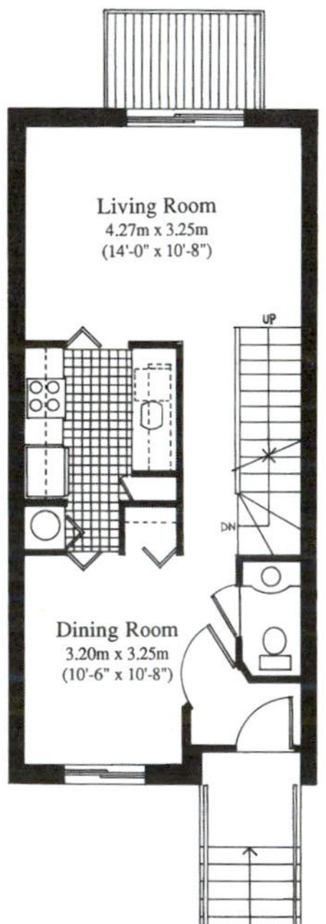

Ground floor with central galley kitchen

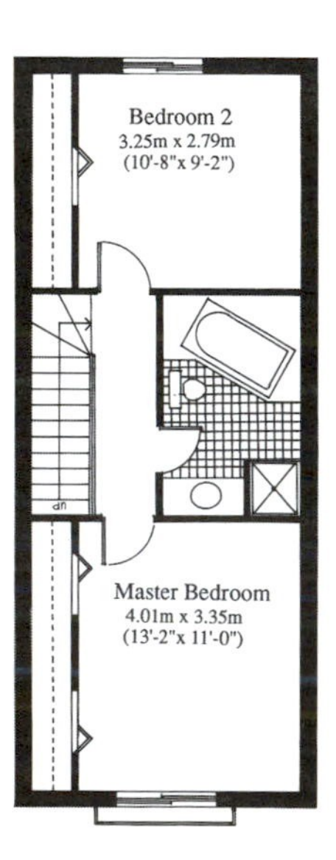

Upper floor with enlarged bathroom (shower and whirlpool bath)

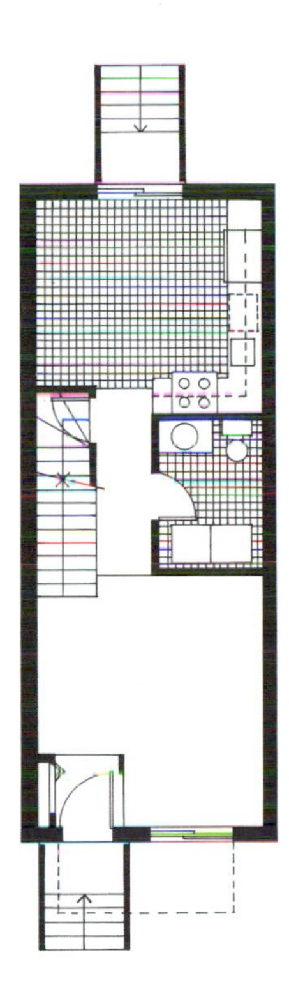

Ground floor with kitchen in the rear and powder room/laundry in the middle

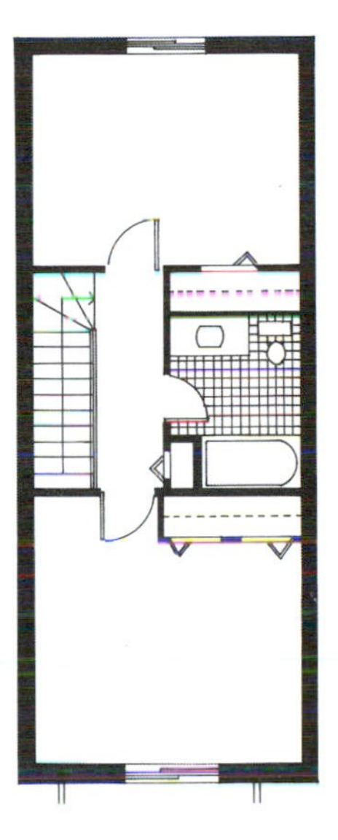

Upper floor with smaller bathroom and larger bedrooms

there was some resistance in one of the infill projects, the complaints faded once the first units were constructed with a quality of exterior finishing equal if not superior to that of the surrounding homes. That change of tone demonstrates how the residents were not as concerned with the price of the units as they were with the image of the project as a whole.

This same principle seems to have applied to the buyers, many of whom had been reluctant previously to purchase a condominium. The prospect of ownership took precedence over the form of tenure in their ultimate purchase of a Grow Home. Their initial concerns seem to have been based on the image of the condominium rather than on the associated legal implications. The townhouse – with its yard space and separate entrance at ground level – satisfied the traditional perceptions and expectations of a house and was generally, even happily, accepted as home.

The builders recognized that they were dealing with a new product aimed at a market which was different from their usual experience (all the builders revealed that the Grow Home was their first involvement with affordable housing) and so they adopted appropriate marketing strategies and selling techniques. They used a common-sense approach in dealing with the prospective client. They delivered promotional brochures directly to the homes of local tenants, many of whom were paying approximately $200 less per month for their apartments than they would be paying for a Grow Home (table 5). They placed advertisements in the leisure sections of the local newspapers rather than in the real estate section. Many advertised on local radio stations. The projects were not referred to as “affordable” but as unique opportunities to acquire quality dwellings at

Project	1	2	3	4	5	6	7	Total
Household income	45,000	38,125	49,575	45,000	50,000	44,500	42,916	45,818
Expenses as tenants	554	573	542	502	543	558	495	538
Expenses as owners	710	754	860	980	770	738	760	796
Percentage of income allocated to housing in current dwelling	18.9	23.7	20.8	26.1	18.5	19.9	21.2	20.8

Table 5 Median incomes and monthly carrying costs (1990 $)

good prices. Emphasis was placed on the standard options and the "luxury" items (fireplaces, whirlpool baths, brick exterior, and wood flooring) – much like the list of options used in car sales. The promotional literature omitted the overall dimensions of the units and mentioned instead the exact dimensions of each room. Salesmen also preferred to stress floor area and room-by-room comparisons of the Grow Home with the buyer's current home rather than speak of overall dimensions. They stressed that the unit's size and shape as well as its grouping contributed to a reduction in energy consumption – an attractive feature for cash-strapped first-time home buyers.

The nature of the marketing strategies showed a keen understanding of the new-found niche in the market. The builders realized that people who were used to thinking that home-ownership was beyond their means would not be looking through the real estate section of newspapers nor would they be familiar with calculating monthly carrying costs from a price list of homes. Such people might also not like the idea of living in a house only 4.3m (14 feet) wide – the same response as the builders themselves before they had walked through a built unit.

By the end of the first year, twelve builders had undertaken Grow Home projects in and around Montreal, and they had built about a thousand units. The most successful projects were those that started the earliest and that received the best prices from subcontractors before the competition could catch up. Once many projects were underway, the home buyer was the ultimate gainer, benefiting from competitive pricing and the increased value of the homes. Even by the year 2000, with approximately ten thousand Grow Home units built in Canada at prices ranging from $69,000 to $95,000, the potential buyer in the market for an affordable single-family home could count on the Grow Home as a steady and dependable housing prototype.

The process by which these projects were realized is characterized by a somewhat unusual and eccentric arrangement of resources. The design was developed in an academic institution with the help of a corporate research grant. The private sector furnished the capital and management tools necessary to see the project through its implementation, and three levels of government provided separate financial incentives for the prospective home buyers. Although each of the parties involved acted independently and for different motives, the product of their combined efforts was the achievement of a common goal: affordable housing.

5 BUYING THEIR FIRST HOMES

Almost nine out of ten Grow Home residents were first-time home buyers who had formerly lived in apartments as tenants. Many were experienced renters, having moved out of their parents' home to rent as students or share an apartment with a friend and later to move in with a spouse. Prior to buying their first homes, some had lived in as many as five rented apartments. The renting routine had been fairly simple. They knew how much they were able and willing to spend. They would have a favourite neighbourhood in mind, close to their work, friends, or leisure activities. The weekend paper would be the main route to finding a place. Then came the usual tour: meeting the future landlord, climbing the stairs, hearing television sounds behind closed doors, trying to imagine who the neighbours were, assessing the options. Once inside the apartment, they would try out the flusher on the toilet and check out the view from a rear window, knowing that they would have to compromise. "This is not really our place," they would say to themselves. "We are here only temporarily." A one-year lease would be signed, at the end of which they would consider either an extension or another move.

Tenants rarely undertake and pay for extensive renovations of an apartment. They will not demolish and rebuild walls and renovate bathrooms. They might paint, install new curtains and hang favourite prints, but they will not replace a window or plant a tree in the backyard. They pay the landlord to do these things. An emotional attachment to a rented unit may exist but rarely the will to make it one's own. The selection of rental accommodation does not involve critical decisions: there are no sacrifices or trade-offs. If a tenant cannot stand the noise through the wall from a neighbour or is troubled by an increase in neighbourhood crime or cannot afford the high rent, the options are very simple: pack up and move, look for another place. If people have to move in the middle of a lease (for, say, a job offer in another town), they can always pay a penalty, sublet the apartment to someone else, and leave.

The process of buying a home is far more complex. Decisions are not as easily reversible. If a person loses a job, separates, or gets a job elsewhere, an owned unit can always be resold, but with a greater element of risk involved: the market may be in a downturn and buyers are difficult to find, the desired selling price may be impossible to obtain, and the seller will have to take a loss. Buying a house involves a firm commitment and a long-term decision, and it bears consequences. The purchase itself is influenced by a wide range of factors, including the characteristics of the individual person or household buying the home, the neighbourhood, the physical features of the unit, and even the social bonding between the household members and the new community (Kaynak and Stevenson 1981). The move to buy a home places a household at a critical crossroads. Financial priorities have to be reconsidered. Accustomed lifestyles may have to be overhauled. Family planning may also have to shift. First-time home buyers are beset by numerous hesitations as they realize that some of their decisions will influence the rest of their lives. The whole process is even harder for buyers with fewer means, who must balance the reality of their financial situation with their aspirations to have a home of their own. Many implications have to be evaluated carefully: buying a small home, for instance, may mean having a smaller family or giving up space devoted to a hobby. The real effects of such compromises may not be known until those who made them have moved into their new home and community and experienced them for the first time.

Renters are not forced to make the same decisions as buyers. Complacency can settle in, and renters accept the fact that they may never own. When the desire to own a home grows stronger, the assumption of trade-offs will begin. The urge to own may occur at a time of personal pressure. A sudden realization that they have not built up any equity or that their retirement prospects look bleak sometimes serves as a wake-up call for renters. An increase in family size may also be a strong motivator to explore the real-estate section of the newspaper. As the biological clock ticks louder, discussions about having a family can force renters to look around their apartment and wonder where the crib would go or if there is enough space in the dining area for a high chair. Once a child – or children – are born, the pressure is on. Decisions are often rushed and many trade-offs made.

Grow Home buyers were under pressure to buy: this is the only explanation for the surprising circumstances that accompanied so many of the purchases. The buying patterns that emerged were characterized by short, immediate decisions, often taken without even having seen the finished product. They had noticed an advertisement for a project in the paper or had closely examined a flyer shoved into their mailboxes. Contrary to Montreal custom, they drove to sites that had no model units and obtained ideas about their future homes from a developer's plan. Three-fifths of the buyers were not actively looking for a house at the time of purchase and seem to have come across the projects by finding out about them in the media or seeing advertisements in newspapers. Almost one-third of the buyers visited only one project, and three in five visited no other project after seeing the one in which they purchased a unit. Two in five decided to buy the unit immediately, while almost a half of all buyers made their decisions within two weeks. The vast majority of respondents (over three-quarters) purchased the house without having seen a model unit.

While the lack of a thorough search has been found to be a major characteristic of house-hunting behaviour, the brevity of the search and the high rate of sales displayed in the Grow Home projects remain unusual. In a study conducted in the Toronto area, Barrett (1976) found that once the decision to buy was made, prospective home buyers engaged in only a brief and casual survey of alternative houses. In a sample of 380 house movers, one-seventh of the total visited one house only, while one-quarter visited two to four houses. Two-fifths of the movers spent less than one month searching. (No correlation was made between the intensity of search behaviour and the socioeconomic characteristics of the home buyers.)

Knowing the dilemmas faced by renters made it heartwarming to see the many newspaper ads announcing monthly amounts for owned units. It was even more exciting to watch Grow Home developments pop up in and around Montreal. Large cement mixers poured foundations into freshly excavated sites. Framers raised stud walls and hammered plywood. These communities had a palpable sense of newness. The roads had not yet been paved and city trees were not yet planted. In the summer of 1991, the year following the demonstration on campus, I drove through these streets. Grow Homes were grouped in rows of three, four, and eight units. I watched a large moving van unload furniture while a father called to his toddler to clear away his toys from the front steps. I saw a woman watering her newly planted shrubs. I noticed the owner of another unit preparing to do some work himself by unloading two-by-fours and drywall from the back of his minivan. I stopped many times to talk with these new owners about their decisions to buy, their new homes and future plans. There was a sense of pride in their voices. They frequently used the phrase "my home" as they reviewed the details of their plans for the empty basement or for a vegetable garden in the backyard.

The bustling building activity was unique in the Montreal housing landscape that only a short time earlier had been suffering through a bad recession. I wondered whether these new owners fit the images that I had envisioned earlier. What was their demographic make-up? What kind of trade-offs had they made to become homeowners? How did it feel to own? How was it to live in a narrow unit and be part of a row? What about the small backyard? Were they happy with their purchase? I knew that they had lived in their new homes for only a short while but I assumed that it would be important to document their first impressions. Standing there in the street, looking at the finished units and seeing more construction in the distance, I realized that we had built not only a community but a laboratory. A study would provide valuable knowledge to be used in future designs. Since the units represented one of a limited number of avenues to home-ownership for this group, we decided to determine whether these units accommodated the buyers' functional requirements and personal preferences or the brisk sales were simply due to the lack of choice in low-rise affordable market housing. We conducted an occupant survey to evaluate resident satisfaction with several aspects of the projects and to find out whether or not there were any trade-offs made with respect to their former dwellings, either willingly or for lack of choice (Friedman and Cammalleri 1992a).

Before studying the occupants of the Grow Homes we consulted other housing occupant studies to determine the degree to which the built environment is able to respond to buyers' housing needs, which these studies ultimately found to be dependent on the level of resident satisfaction with the new surroundings, community, and home. Fried and Gleicher (1961) were among the first to suggest that residents' satisfaction might be a more appropriate criterion for evaluating the quality of housing than physical characteristics such as structure and plumbing. Housing, neighbourhood, and community satisfaction have been found to be interrelated (Campbell 1981), suggesting that comprehensive models for evaluating resident satisfaction should always include multi-faceted sets of attributes. Onibokun (1974) defined the relative habitability of housing for public tenants in terms of four interacting subsystems: the tenant, the dwelling, the environment, and the management. Hempel (1976) has included measures in his own assessment model to account for the overall home-searching experience. The importance of previous location and social environment has also been emphasized as a determining factor of satisfaction (Fried and Gleicher 1961), as well as the demographic characteristics of the occupants. For example, certain types of households (younger, married, female heads) were consistently found to express lower levels of satisfaction with any given context due to their different needs, aspirations, and abilities to change their context (Galster and Hesser 1981). Another study confirmed that higher occupant satisfaction levels were linked to home-ownership and single-family dwellings (Rent and Rent 1978).

While the effect of demographic, psychological, and sociological factors on overall housing satisfaction has been well documented, the dwelling unit itself is the dominant factor in any occupant study, followed in importance by the immediate surroundings of the neighbourhood. A 1987 study by Gruber and Shelton supports the notion that house and neighbourhood should be distinguished as separate realms within the residential environment. The results of their study suggest that satisfaction with neighbourhood is unrelated to satisfaction with home, and that the house itself is the most important of the three realms of house, neighbourhood, and community. Rent and Rent (1978) provide statistical and logical justification for evaluating the neighbourhood and the housing unit separately. Ha and Weber (1991) also conclude that housing quality is the most influential component of housing satisfaction. A study by Baillie (1990) finds that in female-headed, single-parent households, dwelling features such as storage and room size are linked to satisfaction more than household characteristics (such as whether the occupant is an owner or renter) or dwelling type.

After reviewing the literature on the subject, we came to the conclusion that since the dwelling unit itself (and the resulting site plan) is the major factor in determining overall occupant satisfaction, the potential for narrow-front rowhousing developments to

Figure 1 Grow Home exteriors in three developments. Clockwise from top: St Laurent; Ste Catherine; and Pointe-aux-Trembles

address the affordability challenge depended on the ability of the smaller, narrower unit to satisfy the buyers' requirements. Our study would assess the occupants' initial reactions to their new physical surroundings and provide immediate feedback to the projects' builders, architects, and planners. Due to the short residency periods, aspects of the project dealing with social interaction and community participation would not have had time to develop adequately and so we did not investigate them in any detail. However, we did make a preliminary assessment of the occupants' overall satisfaction by asking two general questions: How well had the occupants' expectations been fulfilled? Would they recommend their purchase to a friend?

We could not determine whether or not occupant satisfaction levels would change considerably in the following months and years, since there are studies which argue both sides of the question (Baillie 1990; Rent and Rent 1978). We decided that assessing the occupants' first reactions would lead to a more accurate evaluation of how their new dwellings actually compared with their former homes. Follow-up evaluations were planned to monitor any changes in opinion and to complete a longitudinal study aimed at assessing the potential for narrow-front rowhousing developments to provide quality affordable housing in the long term.

We visited the new Grow Home owners in the winter of 1992, approximately nine months after they had occupied their units. We selected seven of the Grow Home projects based on location, advanced level of completion, and consistency in unit design (figure 1). We visited 236 households in these projects and asked the owners to complete a ten-page questionnaire. A total of 196 questionnaires was completed (a response rate of 83 per cent). We also obtained information from the builders through general discussion and structured interviews (Friedman and Cammalleri 1992b).

So who were the buyers? For the most part, the Grow Homes attracted the demographic group for which they were intended (table 1). The households were made up mainly of young couples without children (just under a third of the total) and couples with children (just over a third). The age group of most of these families (twenty-five to thirty-four) was typical of first-time home buyers at the time. Most of these couples had got married and begun families later than earlier generations, and the pressure to buy a home had mounted as they planned or already had their first child. The prospect of frequent moves with kids did not appeal to them as they sought to achieve a more stable lifestyle. Some of the living arrangements were also typical of the time, with one in fourteen households common-law couples (Mandel and Duffy 1995). Two groups of buyers made a striking contribution to the Grow Homes on the affordability front: single-person households accounted for a sixth of all buyers, and single-parent families for a tenth. These two household types were usually renters, since they simply could not afford to buy. Those who did become owners generally bought an apartment rather than a house. The low cost of Grow Homes had enabled single-income earners to own and maintain a home after renting for many years, an uncommon occurrence in the Montreal housing market.

Household Size	(%)
One person	16.0
Two persons	40.7
Three persons	32.0
Four persons	7.2
Five persons	3.1
Six persons	1.0

Household Type	
One adult	16.5
Two or more adults	3.6
Single parent	9.8
Couple	32.5
Couple with children	37.6

Employment Status	
Self-employed	9.5
Full-time	69.6
Part-time	10.8
Unemployed	6.3
Full-time student	0.6
Retired	0.6
Other	0.3

Education	(%)
Primary school	1.7
Secondary school	35.7
CEGEP	20.8
University	30.4
Trade school	11.4

Occupant Age	(%)
0–17	22.1
18–24	12.8
25–34	43.7
35–44	15.2
45–54	4.0
Over 55	2.1

Household Income	
Below $20,000	1.6
$20,000–29,999	5.5
$30,000–39,999	25.1
$40,000–49,999	30.1
Over $50,000	37.7

Number of Income Earners	
One-income household	36.7
Two-income household	58.7

Occupation	
Farming, fishing, forestry, logging, mining and quarrying	0.3
Labourer	3.2
Semi-skilled	5.7
Skilled tradesman	18.1
Sales, service, clerical	36.5
Professional	11.1
Managerial	15.6
Homemaker	3.5
Other	6.0

Table 1 Grow Home household characteristics (by percentage of respondents)

Andrée Tardiff was one these renters-turned-owners (figure 2). As a clerk in the suburban municipality of Ville St Laurent, she earned a modest income and had lived alone in a rental unit into her early fifties. She never pictured herself as someone who could afford a suburban home or even as someone who would want to. But one day on her way home from work she noticed a large site sign announcing a new project with homes in the $90,000 range. She stopped to ask for information and picked up a flyer which she later read over dinner. The developer had calculated and printed in bold on the pamphlet the amount of a monthly payment – $825, almost the same as her monthly rent. The decision was a no-brainer. She had money saved for a 10 per cent down-payment. Her bank manager welcomed a dedicated employee with a secure income. In 1991 she became a homeowner.

Another surprising characteristic came to light when we asked for buyers' incomes. Most of the households (almost two-thirds) had incomes below $50,000, leaving over a third with higher incomes. Despite the fact that they could afford a larger and more expensive house, these households with incomes over $50,000 opted to buy a smaller unit such as the Grow Home. Calculations based on the occupants' carrying costs revealed that they were allocating between 18.5 per cent and 26.1 per cent of their incomes on housing. They had realized the simple wisdom of not spending the allowable maximum of 32 per cent on shelter expenses. Other housing costs had risen in recent years. Many two-income households needed a second car, and some did not want to give up the family trip or outings to restaurants because they had bought a home. By making the

Figure 2 Grow Home buyers
Michel Cérallo (a bookkeeper) and his companion Sophie Cailloux (a teacher) are typical Grow Home buyers: in their late twenties, fresh out of university, they rented for a while before buying their unit in Laval for $86,000, then made some changes – a fireplace in the living room and a finished basement.

Andrée Tardiff (an information clerk in a municipal office) was an apartment tenant for twenty years before deciding to buy her $93,000 unit in Laval; she paid a 10 per cent down-payment and her monthly payments are $825, only slightly higher than her previous monthly rental expenses.

trade-off of buying a small home, they were able to hold onto most things they wanted. Michel Cérallo (an accountant) and Sophie Cailloux (a teacher) bought a Grow Home with plush wall-to-wall carpeting, a glass-encased fireplace, and ceramic tile in the kitchen and bathroom, pushing the final price of their unit up to $95,000 (figure 2). After paying down $15,000, they were left with monthly mortgage payments of only $675, leaving them with cash for other non-housing pursuits.

When buyers were questioned on the reasons for purchasing their Grow Homes, almost a half of all respondents reported price to be an attractive feature, while nine in ten buyers rated price as "extremely important" in their decision to purchase. The rapid rate of sales, coupled with the importance of selling price reported by the buyers, indicated that the projects filled a market void. In fact, housing data acquired in the year of the campus demonstration indicated that four-fifths of all renters between the ages of twenty and forty-four did not have the means to buy a starter home in Montreal (CMHC 1991). Only one-tenth of the available housing supply was affordable to this group.

We came upon another important key to affordable home-ownership when we compared the buyers' monthly expenses as tenants with their new expenses as owners. In most cases homeowners paid $200 more than the monthly amounts they paid as renters. The trade-off was that they bought a small unit. In all the surveyed projects that additional amount of $200 became a benchmark of what those who had decided to switch from tenant to owner would pay. Many pushed themselves to the limit. In almost three-fifths of all households two incomes were needed to pay the mortgage, despite the presence of young children. No long pause without paid work had been available after the birth of the children, and the need to hang onto the second income was essential.

Working close to home is also critical in the lives of first-time buyers. It is one of the main criteria in the selection of their new home's location. The decision of where to buy a home involves a range of attributes, including social status, ethnicity, school-age children, and proximity to friends and relatives. Location can influence lifestyle and even the daily conduct of many young households. For many families, the day has become a race against time. A parent wakes up early, often before sunrise, and gently wakes a sleeping toddler, praying that the kid will be in a good mood, not have a cold, and be kindly disposed to the breakfast cereal. While the toddler eats, the rush continues: a quick load of laundry, a gathering of materials for the day, a scan of the fridge to plan dinner. Then it's into the snowsuit with the kid, out to the car, dash back indoors for a favourite toy, outside again, clear the snow off the windshield, and off to the daycare. From there it's straight to the expressway, hoping there won't be a traffic jam to delay the 8:30 appointment at the office. The rush to beat the afternoon traffic is no different: dash to make the daycare on time, stop at the supermarket, cook, eat, prepare for the next day, and fall asleep in front of the TV.

Distance to work and proximity to amenities (daycare, medical, and shopping facilities) have become crucial. Extended families used to provide a helping hand; contemporary life and mobility have changed that. For most Grow Home buyers, the trade-offs began with location. Living close to the city centre in an established community became too expensive. Affordable homes were built where inexpensive land was available – far from the centre. Transportation routes expanded along with urban sprawl, making a drive to work of one or two hours not uncommon. Life has become very dependent on the car, which has almost become a family member. Ensuring that a newly purchased home comes with two parking spots has become part of the shopping list. This pattern has meant that there has been no justification for any increase in bus transportation frequency. It was a clear trade-off: assume a certain lifestyle and drive longer in return for home-ownership.

Moving to distant locations coincided with another image: the suburban ideal. When polled on their general preferences, the buyers in our study indicated that they had very much wanted to live in a suburban setting in a new, single-family detached dwelling (table 2). Just over a half of all respondents rated the suburbs as their most preferred location (regardless of price).

When the buyers compared their new Grow Homes with their former dwellings, they indicated marginal improvements over their previous homes in two location characteristics (proximity to schools and medical services) and deterioration in the remaining three (table 3). The overall satisfaction of the occupants with their new location was not reflected in the comparison ratings. Those characteristics which were considered to be improvements were also not rated as highly as their original preferences. The greatest compromise for the occupants was their distance from work – a significant finding when we realized that they had ranked this highest in their list of preferences. Public transportation was the second highest source of compromise, with a third of respondents rating it as worse than at their former dwelling.

Yet the high preference rating assigned to the suburban locations indicated that locating projects off the island of Montreal did not cause major problems for buyers. (The Grow Home developments on the edges of the island of Montreal or in areas off the island involve commuting distances to the downtown area ranging from 15 to 40 kilometres [9.4 to 25 miles].) Access to services that are not part of a daily routine such as medical services and shopping facilities was of little importance to the majority of occupants. With almost all of the households owning a car and close to half owning two cars, access to these services did not seem to be a critical issue. Proximity to schools may have not been considered important due to the availability of school bus service. The distance from work, however, represented an important compromise for most of the occupants. We therefore examined transportation to and from work in greater detail and found that there were increases both in general car use as a primary mode of transportation and in commuting time. We also found decreases in both public transit use and in walking. Given the very high priority assigned to proximity to work, the equally high level of compromise that was demonstrated in this regard, and the reasonable degree of expressed satisfaction, we concluded that there was a strong willingness on the part of the buyers to make a trade-off in their proximity to work in exchange for home-ownership.

Once home buyers had made their choice of location, they drove to see a model unit of a Grow Home.

Preferred Number of Bedrooms[1]	(%)
One bedroom	0.5
Two bedrooms	54.6
Three bedrooms	43.8
Four bedrooms	0.5
More than four bedrooms	0.5
Specifically Looking for a Newly Built House	58.5

Important Features Desired in a New Home[2]	
Second/big bathroom	7.7
Sufficient storage	6.7
Natural lighting	6.4
Basement	5.8
Garage	4.3
Second/big bedrooms	4.0
Open plan	4.0
Sufficient amount of total space	3.7
Functional kitchen	3.7
Two storeys	3.7
Private parking	3.4
Quality of interior finishes	3.1
Good location/neighbourhood	3.2

Importance of Location Characteristics[3]					(%)
Proximity to:	1	2	3	4	5
Work	5.2	2.6	17.7	27.6	46.9
Public transit	23.3	5.8	12.1	20.1	38.6
Medical services	14.2	17.4	32.1	22.1	14.2
Shopping facilities	15.3	13.8	39.7	18.5	12.7
Schools	31.1	12.6	11.5	19.7	25.1

Preferred Location[4]	1	2	3	4	5
City centre	77.6	13.1	2.2	4.4	2.7
10 mins. from centre	11.8	35.8	18.7	13.4	20.3
Suburb	3.7	3.7	19.8	20.9	51.9
Small town	7.2	14.4	27.6	41.4	9.4
Country	23.9	17.9	21.2	14.2	22.8

Preferred Type of Dwelling[5]	1	2	3	4
Single-family detached	12.3	5.3	4.3	78.1
Semi-detached	11.1	22.1	63.5	3.3
Townhouse	4.6	56.8	24.4	14.2
Condominium apartment	77.0	10.4	6.3	6.3

1 Based on percentage of respondents
2 Most frequently mentioned attributes; percentage based on 327 entries
3 Percentage of respondents; scale of 1 (not at all important) to 5 (extremely important)
4 Percentage of respondents; scale of 1 (least preferred) to 5 (most preferred)
5 Percentage of respondents; scale of 1 (least preferred) to 4 (most preferred)

Table 2 Grow Home buyers' preferences and desired features

Location	Sat.[1]	Comp.[2]
Proximity to:		
Work	3.736	2.851
Public transit	3.615	2.892
Medical services	3.718	3.138
Shopping facilities	3.766	3.196
Schools	3.766	3.196
Overall average	**3.697**	

Site	Sat.[1]	Comp.[2]
General image	4.042	3.876
Sense of community	3.792	3.665
Safety of children	3.785	3.739
Location of parking	3.540	3.584
Size of front yard	3.346	3.577
Size of backyard	2.937	3.553
Level of privacy	3.219	3.483
Overall average	**3.513**	

Unit		
Overall design/layout	4.152	3.949
Total amount of space	4.169	3.778
Amount of storage space	3.95	3.808
Flexibity of space	3.689	3.630
Exterior appearance	4.193	3.862
Interior appearance	4.218	3.938
Quality of exterior finishes	4.069	3.740
Quality of interior finishes	4.193	3.483
Amount of natural lighting	4.098	3.542
Amount of cross-ventilation	4.038	3.652
Soundproofing between units	3.500	3.657
Overall average	**3.933**	

Interior Spaces		
Entrance	3.236	3.243
Living room	3.844	3.489
Dining room	4.060	3.648
Kitchen	3.926	3.598
Ground-floor bathroom	4.178	3.827
Master bedroom	4.487	3.823
Second bedroom	4.373	3.755
Upper-floor bathroom	4.417	4.076
Hallway	4.126	3.660
Stairs	4.254	4.000
Overall average	**4.092**	

1 Satisfaction: average score on a scale of 1 (very disappointed) to 5 (very satisfied)
2 Comparison: average score on a scale of 1 (much worse than former dwelling) to 5 (much better than former dwelling)

Table 3 Occupants' level of satisfaction and comparison of Grow Homes with former dwellings

They found a trailer instead, with plans pinned up on the walls. Buying a home is often a process of seeing a place and matching it with desired living conditions and aspirations. Grow Home buyers clearly would have liked to buy a single-family detached home. In fact, over three-quarters of all buyers would have preferred to live in such a unit. The rowhouse was a trade-off for them. The Grow Home represented a step down from their first choice. They still got a backyard but they were in attached houses, as part of a row. However, the townhouse, with its separate entrance at ground level and private yard space, appeared to satisfy many of the traditional perceptions of what a house should be. But the tendency to accept trade-offs did drop significantly when any of the traditional elements of home-ownership were compromised. This became evident with the relatively high levels of disappointment with the size of the backyard area – a compromise of expectations in terms of territory and privacy (figure 3). These observations are supported by the findings of a post-occupancy evaluation by Kantrowitz and Nordhaus (1980), where territoriality, site security, and private outdoor space were the most consistent issues arising out of the analysis of site design. The lack of adequate yard space in some cases may have also reduced the perceived distance between neighbours, a notion that has been correlated with lower levels of neighbourhood satisfaction (Lansing and Marans 1969). It is worthwhile noting that fences were not provided by the builders and had to be installed by the residents themselves. During the summer of 1991 there was no clear demarcation between lots, and as a result there was a feeling of a lack of privacy. Fences were put up in most of the projects the following summer.

In general, the aspects that represented the greatest improvement over the occupants' former dwellings were also those that generated the highest levels of satisfaction, supporting the notion that occupants' satisfaction with their new homes was related to their former housing situation. The occupants were happy with most of the site attributes, particularly with the projects' general image, sense of community, and the safety of children. The exterior appearance of the homes was also well liked, and a large majority indicated that their housing situation in general was positive. These tendencies are in line with the findings of previous studies (Gruber and Shelton 1987; Wiedemann and Anderson 1982) which demonstrate that occupant satisfaction with the neighbourhood realm appears to be more aesthetic than substantive, based on perceptions of attractiveness, friendliness, and safety.

With regard to the site plan, some high levels of improvement were accompanied by relatively low levels of satisfaction (figure 4). For instance, many of the new Grow Home occupants had only balconies for outdoor space in their previous dwellings and most had no backyard access at all. Although the satisfaction ratings for the size of backyards, front yards, and privacy all represented an improvement over the occupants' former homes, the overall (non-comparative) rating was still lower for these features than for others. Such results indicate trade-offs that were made unwillingly.

The apparent difficulty experienced in the accommodation of parking in most projects resulted in reduced yard space and, in some cases, excessive paving. The fact that the units were sold as condominiums implied that there were no legal boundaries

Figure 3 Grow Home backyards

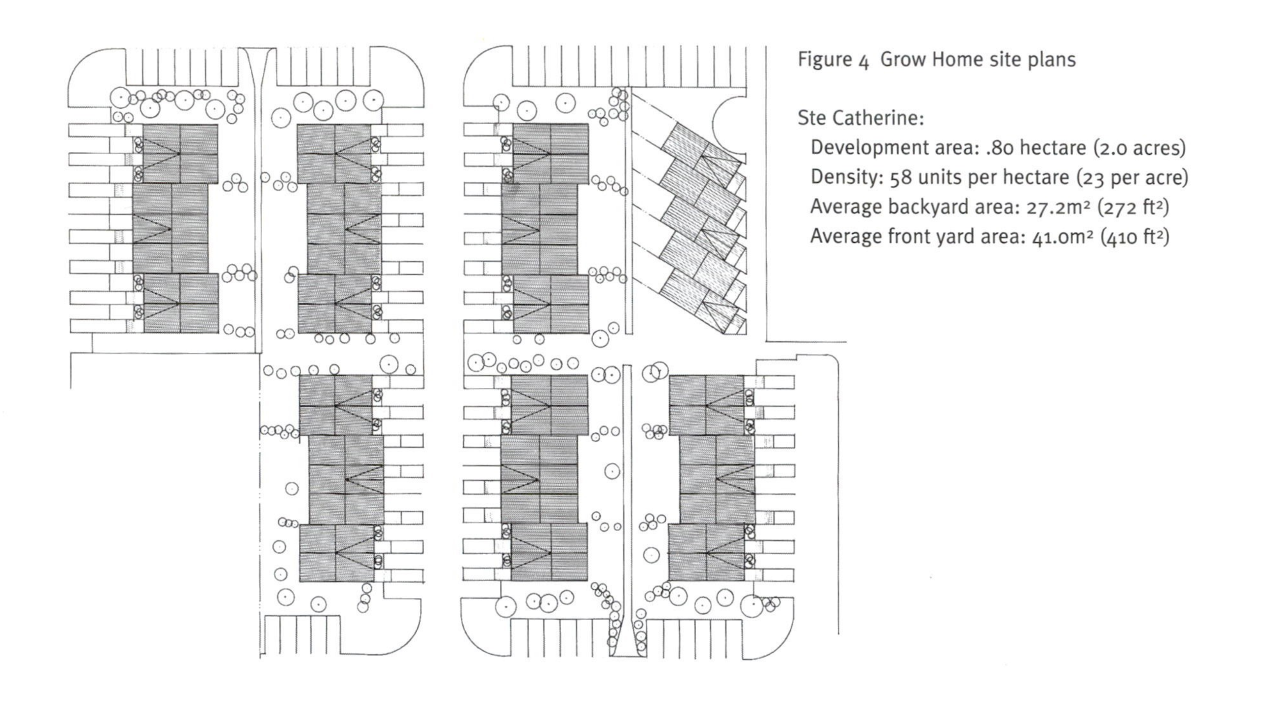

Figure 4 Grow Home site plans

Ste Catherine:
Development area: .80 hectare (2.0 acres)
Density: 58 units per hectare (23 per acre)
Average backyard area: 27.2m² (272 ft²)
Average front yard area: 41.0m² (410 ft²)

Chomedey:
Development area: .28 hectare (.7 acre)
Density: 43 units per hectare (17 per acre)
Average backyard area: 46.3m² (463 ft²)
Average front yard area: 28.1m² (281 ft²)

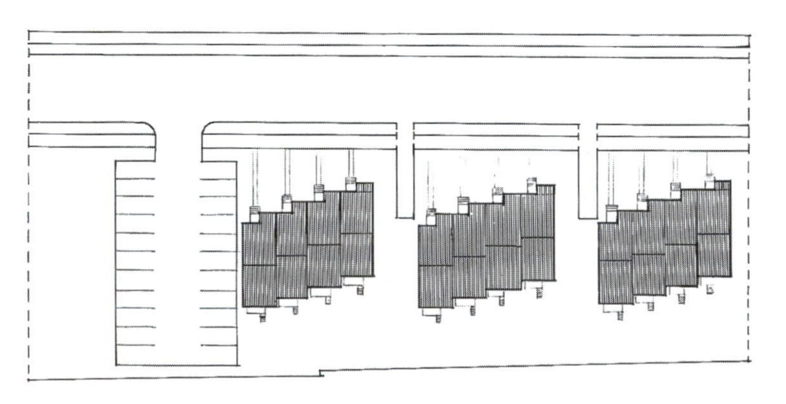

Ste Dorothée:

Development area: .58 hectare (1.5 acres)

Density: 45 units per hectare (18 per acre)

Average backyard area: 45.9m² (459 ft²)

Average front yard area: 41.2m² (412 ft²)

Longueuil:

Development area: 1.71 hectare (4.3 acres)

Density: 41 units per hectare (16 per acre)

Average backyard area: 56.9m² (569 ft²)

Average front yard area: 30.4m² (304 ft²)

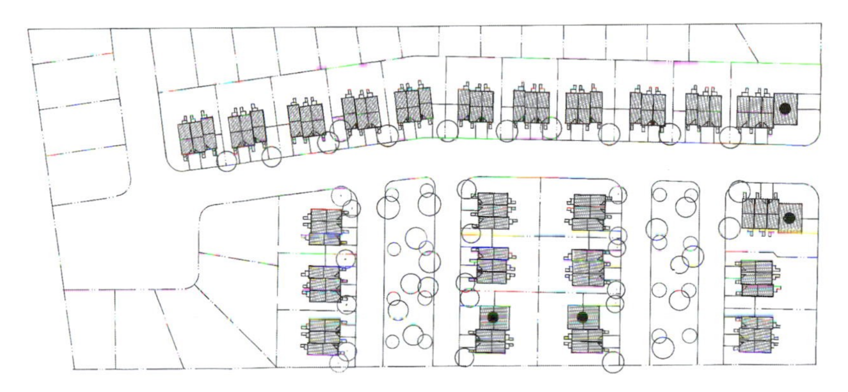

to define the occupants' property limits. We therefore examined the issue of private outdoor space, both for leisure and parking. The occupants were asked whether or not they had experienced any problems in sharing this yard space with their neighbours and, if so, what type of problems occurred. Less than a fifth of the respondents answered that they had experienced any difficulties. The most prominent problem with the backyard area was the lack of privacy and disagreements on types, sizes, and locations of fences.

Once Grow Home buyers had made up their minds about the neighbourhood in which they wanted to live, they began to examine the home itself. Based on past experiences, cultural background, and personal aspirations, buyers develop an image of a desirable home. As they walk through a model home or examine a plan, they match what they see with their idealized image. One household member will closely scrutinize the kitchen while another rushes to the master bedroom. As they walk through the rooms, they evaluate the suitability of the space to their own needs. They try to picture future life stages in the home. Which room will be for the baby? Will the dining room be big enough for a family gathering? They make a decision to purchase only when they have made a match between aspirations and reality.

When our young householders first saw the Grow Home, they liked several of its features. They welcomed the townhouse aspect, the stacked floors, the layout. They could clearly divide public and private activities. When friends or relatives came over, they would not see the mess upstairs. They liked the downstairs which contained everything they wanted – in small portions. There was a small entryway, a decent-sized living room overlooking the street, a dining room in the middle, and a kitchen in the rear. There was even a small washroom. The lower floor did not look at all cramped. They loved the upstairs: a fair-sized master bedroom, a smaller child's room, and a bathroom in between. And of course there was all that extra space in the basement. In the lifestyle that they envisioned, there was not much to take care of: the house was not very big but at least it was maintenance free.

The unit characteristics, particularly the interior spaces of the homes, were clearly the most satisfying aspect for the buyers (figure 5). A substantial number reported that they were very satisfied with seven aspects in particular: interior appearance, exterior appearance, quality of exterior finishes, total amount of space, overall design/layout, amount of natural light, and amount of cross-ventilation. The lowest satisfaction ratings for the homes were reserved for the quality of interior finishes and the soundproofing between units. We studied the issue of acoustic privacy by evaluating sound transmission through the common walls. Although three-quarters of respondents reported being able to hear their neighbours, only a tenth of all occupants felt that the noise was in some way disturbing. The apparent willingness of the occupants to deal with occasional sound transference from the neighbours was most likely due to their background as tenants. Most of the respondents had lived in apartments, and the level of acoustic privacy in their new homes represented an improvement over former living conditions.

The Grow Home occupants graded their unit characteristics and interior spaces positively, considering

Figure 5 Grow Home interiors, Laval, Quebec

them to be an improvement over their former circumstances. Interior and exterior appearances were rated among the top three sources of both improvement and satisfaction, while the quality of interior finishes offered the weakest improvement over the former dwelling and had the lowest level of satisfaction. The units themselves accommodated most of the buyers' preferences since they were of new construction, possessed a large second bathroom and two finished bedrooms, and held out the possibility of adding a third bedroom in the unfinished basement. The only substantial source of compromise appears to have been with the preferred dwelling type, since the vast majority of respondents had not rated the townhouse as their favourite type of housing. There was no evidence of compromise, however, with any of the unit attributes or with the interior spaces. With two floors of living space (which could be extended into the basement), a separate private entrance at ground level, and the provision of private outdoor spaces for leisure and parking, the small size and narrow configuration of the units did not appear to inhibit any of the buyers' expressed needs or functional requirements.

Grow Home buyers made the transition from renting to owning. They made trade-offs in order to become owners, location being the biggest one for many. But in return they had a home they liked. As I drove away from the site back in 1991, I thought that it would be interesting to return to visit them again in a few more years and find out what had become of their lives in the Grow Home.

6 MAKING IT THEIR OWN

The first years were busy for the new Grow Home owners. The reality of home-ownership and the consequences of the trade-offs that some had made were a daily presence. Soon after they had unloaded their furniture and eaten their first meals in the new dining room, the residents set about making the units their own. It took time to get accustomed to the new homes: the sounds that came from the street, the best exposure in summer for a plant. Living in the new homes was a process of getting used to volumes and sizes. The owners tried to fit their old furniture into the new spaces and found themselves moving the chairs and sofas around in an effort to achieve a satisfying arrangement. They also turned their attention outdoors. There was no landlord to plant trees and fix up the yard. They finally got to go to the flower market in the spring to buy their favourite plants and experiment with colours and then stand outside watering it all each evening. There was also the process of creating a community. New neighbours lived in the same row of homes, and it would take a while to get to know them well enough to ask for a cup of sugar or to invite a kid the same age as their own to come over and play. There were common concerns, casually broached while seeing one another on the front porch: the efficiency of garbage collection, trees not yet planted by the city.

Life carried on for the Grow Home owners: job promotions and job losses, changes in lifestyle and habits, the need to adjust to hectic daily routines. Family changes occurred, too: singles found spouses who moved in with them, or they sold their units and moved to live elsewhere; single parents remained while grown children moved out to form their own households; babies became toddlers, and toddlers grew into school-age kids; new babies were born. The dynamics that characterized the lives of the occupants also left an imprint on their homes. New demands were placed on existing spaces, furniture had to be rearranged yet again, new functions were created that had not been included in the original plan. Babies' highchairs were unloaded in some homes while cribs were assembled in others. Suddenly play areas were needed in the backyard. The homes had taken on lives of their own.

I returned to visit these communities many times. I accompanied colleagues from other universities who read about the Grow Home in academic journals, builders who were interested in building them in other places, and my own students on field trips. Sometimes as I paused in front of a row, somebody would step out from one of the homes. Introducing myself, I would engage householders in conversation: Who were they and where did they live before? Did they like the community and the home? I asked them about changes they had made in their homes: Did they work on the unfinished space? Did they make changes in other parts of the house? Who exactly did the work? The conversations were fascinating, and they made me realize yet again that housing is less about architecture than it is about people and their lives and aspirations. The home is the place where daily events are strung together to become larger experiences. I wanted to return to these new Grow Homes and get a sense of the big picture while gathering together the smaller images. I would document the process of how the new householders made the homes their own, and I would describe the outcome.

When one of my students in the Affordable Homes Program, Aurea Rios, an architect from Panama, asked about a suitable topic of research for her Master of Architecture thesis, I described the Grow Home communities to her. I suggested that documenting the modifications made by the residents of the homes in the three years after moving in would provide an intriguing insight into the homes and the lives of first-time buyers. Aurea liked the idea. We reflected on the potential components of such a study and began by consulting other studies to determine *when* people make changes to their homes. In their study of plex housing in the Notre-Dame-de-Grâce neighbourhood of Montreal, Teasdale and Wexler (1993) identified two types of life events that can act as motivators to renovate. The first type corresponds with developmental stages in the life of the family and is comprised of birth, the passages from pre-school to childhood, from childhood to adolescence, and from adolescence to early adulthood, the departure of children from the family home, preparations for the empty nest, and caring for aged parents. The second type of life events includes divorce, widowhood, family fusion, and women's entry or return to the paid work force or to school and unemployment. We knew that most Grow Home residents were at the stage of early family formation. For the majority, we assumed, the birth of children and their growing up would be the primary reasons for making modifications.

We were also curious to find out whether or not the home itself as a vehicle for self-expression played a role in the process. We suspected – and confirmed from studies on residential adaptation as a natural activity of homeowners – that making modifications was a means of matching the home to personal needs and aspirations as well as a way of expressing individual values. Becker (1977) remarks that the freedom of residents to manipulate their living spaces is important for several reasons: “functional requirements, the need for change and variety, the ability to express individuality, and the desire to feel that one has the power to control a piece of the world.”

Morris and Winter (1975, 1978) maintain that a gap exists between actual housing conditions and those prescribed by culturally derived norms when a house does not meet the residents’ standards. They also note that the appearance of this gap in a household can easily result in residential dissatisfaction. When this dissatisfaction becomes strong enough, the family tends to make housing adjustments as a means of eliminating the gap. Typical behavioural changes which are made in response to the need for housing adjustments include moving to a different dwelling, changing the family composition to suit the housing (by, for instance, postponing the birth of children), and residential adaptation. This latter response refers to the various activities householders can undertake to make their homes match their needs and can include any kind of physical and spatial changes such as additions and remodelling. These were the changes we were interested in – we wanted to see what the residents had done and understand why they had made those changes. We also wanted to determine how they did the work (particularly with respect to lower-income residents): did they hire a contractor or do the work themselves?

In order to focus our study, we selected the first Grow Home development built by Leo Marcotte in Pointe-aux-Trembles and occupied in the summer of 1991. All the units had identical dimensions and layout. As we had found in an earlier study, the primary age group of the residents was twenty-five to thirty-four years old (45.4 per cent of the total), followed by those thirty-five to forty-four (34.7 per cent). The overwhelming majority of the occupants were French-speaking. We considered a total of 176 households to be suitable for the study. Each home was visited and the residents were asked to complete a questionnaire of twenty-nine questions. In all, 141 questionnaires were collected, representing an 80 per cent response rate. In addition to the questionnaires, twenty-four interviews were conducted. Residents who demonstrated (through the questionnaires) the most interesting modifications were asked for an in-depth interview so we could obtain a comprehensive picture of their work and their reasons for undertaking it.

The space that interested us most was the basement, intentionally left unfinished as a cost-saving measure (figure 1). We were eager to find out whether or not the owners had completed it and if they had, how they had done so (figure 2). It came as no surprise to find that the basement was, in fact, the place in the house where most changes were carried out (12.6 per cent of total work done in the home) since the unpartitioned space offered ample flexibility for making adaptations in contrast to the relative limitations of all the other areas (figure 3). The basement therefore represented the primary growth space in the house. The types of modifications included work on walls, floors, electrical features, ceiling, doors, windows, storage, and plumbing features (figure 4). As anticipated, the basic reason that residents – mainly families with children (64.8 per cent of all households) – made changes to the basement was the need for additional space.

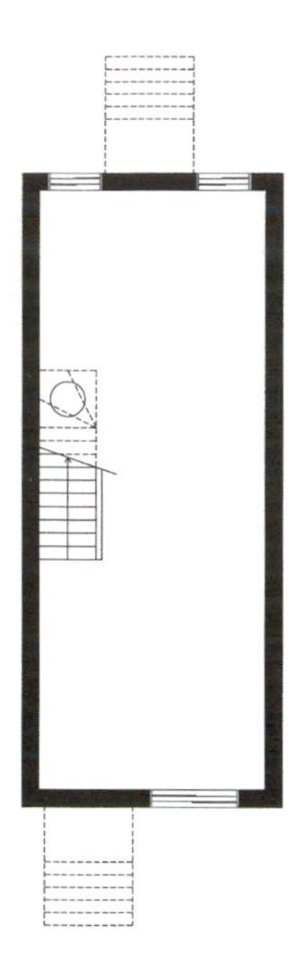

Figure 1 Basement layout plan (before alterations)

The basement was the space that most easily accommodated the needs and desires of the residents, where they could expand their small houses and add missing uses. Of all residents, 61.7 per cent created new spaces in the basement (figure 5). Although a high percentage of residents (87.9 per cent) planned to create new spaces in the basement prior to occupancy, only 62.1 per cent of this group actually undertook what they had planned. Lack of money and time accounted for this gap. Couples with children spent the most on a renovation: an average of $2,829, higher than the average provincial expenditure of $1,188. Single people living alone spent an average of only $623, slightly higher than half of the provincial average (Statistics Canada 1993). It was encouraging to us to find out that every basement was arranged differently – that is, that the design permitted many different options. The main reasons given for creating new rooms in the basement were the need for additional space, the desire to increase the resale value of the house, and the urge to upgrade its appearance.

Of all the work performed in the basement, the creation of a family room received the most attention (25.4 per cent of all work done in the basement) (figure 6). People enjoy having a family room as a place for informal activities such as watching TV and listening to music, where children can play at the same time, leaving the living room as the formal area for receiving visitors. Creating a music room was the dream of Christian Pelchat who bought a Grow Home unit near his place of work after living for many years with his wife in a small one-bedroom apartment in downtown Montreal. He had seen the Grow Home development going up as he travelled to work, and after investigating the project, fell in love with the area and also with the concept. He liked the fact that there was an unfinished basement with the same floor space as each of the floors above it. The high ceiling was an advantage too, giving the impression of a large room. He devoted half of the total basement space to his music room and built the whole room himself, adding acoustic insulation to the ceiling and walls and installing all the electrical connections and speakers.

The need for such spaces in the home is confirmed by Zeisel et al. (1981) who pointed out the need for two separate living areas: an informal sitting room/den and a formal parlour/living room. The location of the family room in the basement has become traditional. As well, some residents also use the fam-

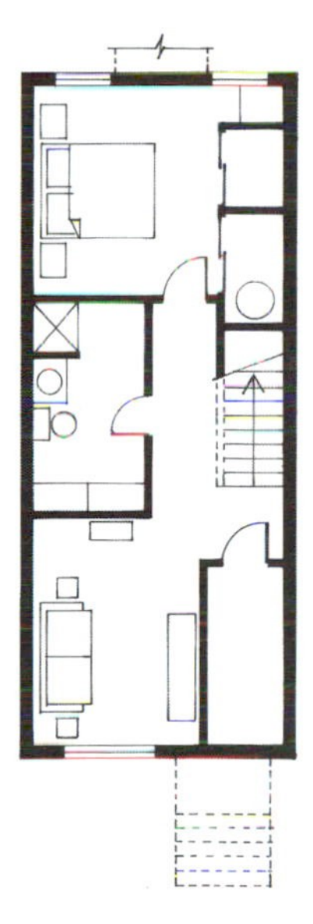

Figure 2 Basement adaptations

Household 1

Household 1 consists of a single person living with a tenant. To make the basement suitable as a separate rental unit and to upgrade the resale value of the house, the owner finished the walls and ceiling with drywall and paint, and built partition walls to create a new bedroom, a full bathroom with washer and dryer, a storage area and a living room. The floor was insulated with plywood and then carpeted, and the stairs were also carpeted. The owner demolished the wall alongside the stairs to make the room appear more spacious. Electrical work included wiring and the installation of lighting fixtures, switches and outlets. Rough plumbing work was done to enable the installation of the shower, toilet, sink and washer. The owner also added a closet and storage area in the bedroom and cabinets in the bathroom and bedroom. All of the new basement spaces received interior doors.

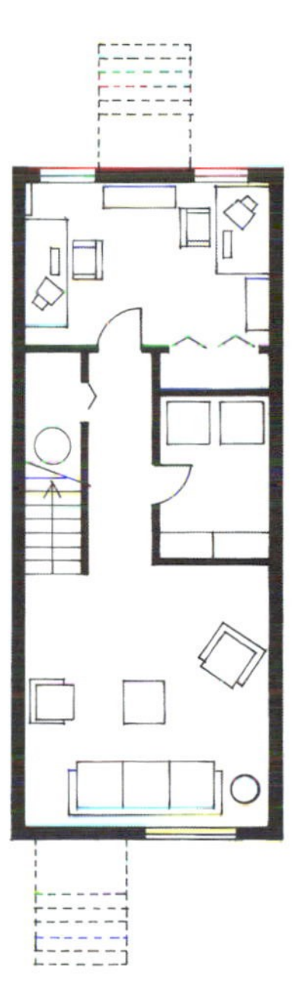

Household 2

Household 2 consists of a common-law couple without children. They finished the walls and ceiling of the basement with drywall and paint, and built partition walls to create a new family room, an office, a laundry room, and a storage space under the stairs. They had originally intended the office to be a bedroom (complete with a closet) but they added some shelves and used it as an office instead. They also added cabinets in the laundry room. To get more light, they created an opening in the wall alongside the stairs. The floor of the family room was insulated with plywood and then carpeted, while the rest of the basement floor was finished with linoleum. The stairs were painted. For security purposes, the owners installed wire netting over the windows. Electrical work included wiring and the installation of lighting fixtures, switches, and outlets. Two interior doors replaced on the ground floor were reused in the basement.

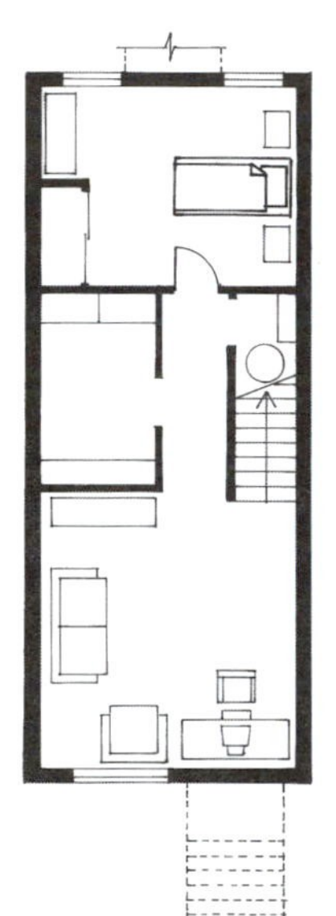

Figure 2 Basement adaptations
Household 3
Household 3 consists of a common-law couple with two young adult children. They built partition walls to create a new bedroom, a laundry room, a storage area, and a family room. At the time of the interview, the walls of the bedroom had been finished (drywall and paint) but the rest of the basement was still in the process of being finished (the drywall ceiling had not yet been painted). The owners carpeted the bedroom floor and simply painted the floor in all the other areas; they intended at a later date to finish the floors of the family room and hallway with wood and the laundry room with tile. Electrical work included wiring and the installation of lighting fixtures, switches, and outlets. Interior doors were provided for all the new spaces as well as shelves in the laundry room and storage area and a closet with sliding doors in the bedroom. The owners put a desk with a computer in the family room which doubles as a home office.

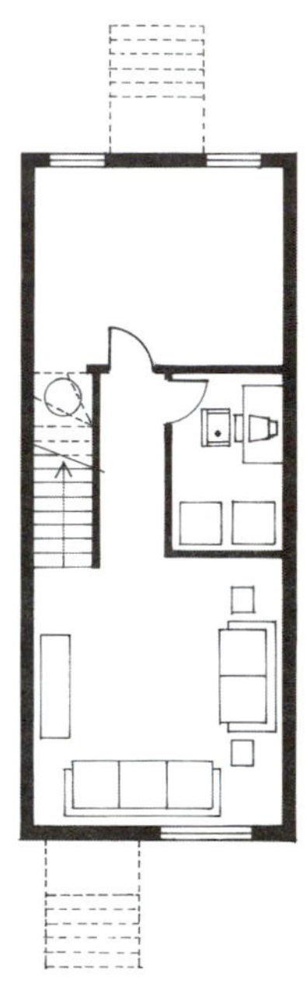

Household 4
Household 4 consists of a married couple with a baby. They finished the walls and ceiling of the basement with drywall and paint, and built partition walls to create a new combination work place/storage room, a combination laundry room/home office (complete with washer and dryer as well as a desk and computer), and a family room. To get more light and to improve the appearance of the basement, they created an opening in the wall alongside the stairs. They carpeted the family room and hallway, and left the rest of the floor bare, to be painted at a later date. For security purposes, the owners installed wire netting over the windows. Electrical work included wiring and the installation of lighting fixtures, switches, and outlets. New interior doors were supplied for all the new rooms.

ily room as a guest room, which makes the basement location suitable since it provides some measure of privacy. The creation of a family room is simple, involving only surface treatment, and does not require a substantial investment of money, work, or time, nor does it cause a major disturbance in everyday life on the upper floors.

Laundry rooms accounted for 22.8 per cent of the spaces created in the basement. The construction of a laundry room can be explained by the fact that many of the families had young children and so a great deal of clothing to be washed. Although the existing plumbing connection was an influential factor in the decision to create a laundry room in the basement, some residents did not locate the laundry room where the plumbing connection was installed, because the original installation did not match their other arrangements for the space. For many, it represented a true opportunity to design their homes, to exercise their personal wishes.

Of the new spaces created in the basement, 10.9 per cent of all work on that level involved storage rooms. In small houses, families with children find that the provided storage space is never enough. In fact, 72 per cent of the residents who made storage rooms in their basements were families with children. Zeisel et al. (1981) remark that "housing residents always feel cramped, no matter how much space they have. They need more space for storage as their possessions grow. They need more space for a play room as their family grows. They need more space to get unsightly utilitarian objects like washers and dryers out of the way." For residents (especially those who live in small houses), it is extremely important to have enough space to store possessions without using spaces intended for other functions. Beck and Teasdale (1977) remark that "basements relieve the storage problem and allow residents scope for improvement and personalization." The most popular type of storage work done in the basement was the installation of cabinets (43.2 per cent). Residents created spaces for additional storage space for items such as food, toys, firewood, tools, machines, and seasonal clothes and articles. Many used imaginative methods and locations for storage spaces, such as underneath the basement stairs. They went through a process of "packing" their homes, using all spaces efficiently and carefully as they made their modifications.

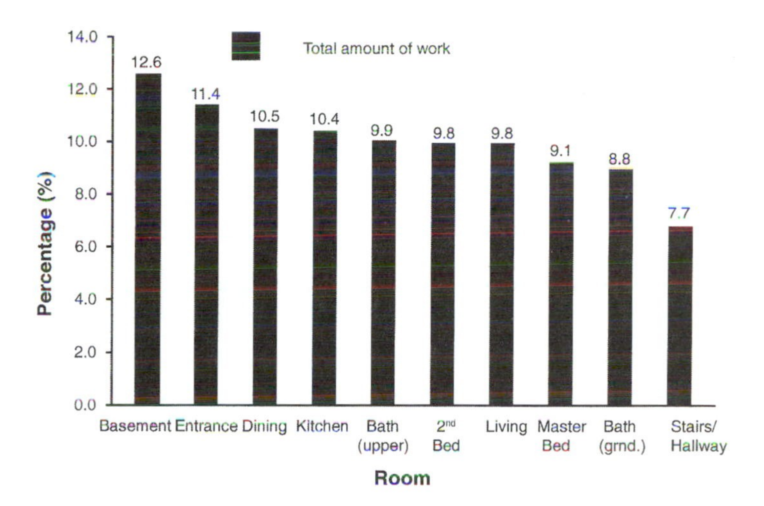

Figure 3 Homeowner modifications, by room

François Laviolette and his wife bought a Grow Home unit in the Cité Jardin Fonteneau development in east-end Montreal, designed by architects Cardinal

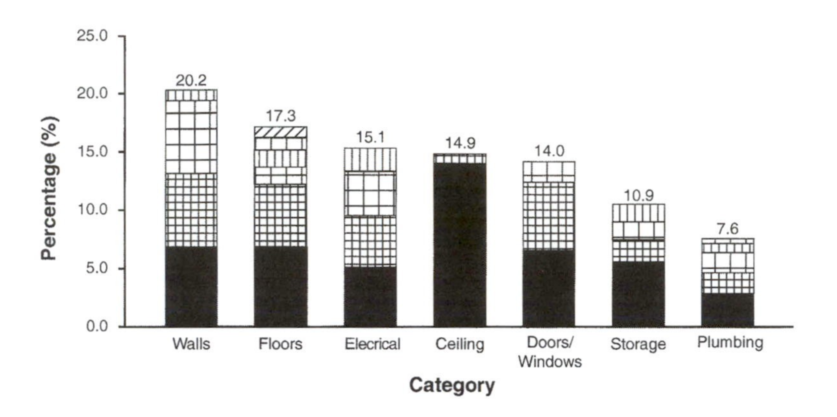

		cem.-tiles 5.7					
	tiles 4.1	wood-carpet 7.6					shower 7.8
	woodwork 6.0	painting 10.5	outlets 16.1			shelves 17.0	toilet 11.8
	partitions 26.5	linoleum 10.5	lighting 25.1		wire netting 16.0	closets 17.0	sink 24.5
	drywalling 29.8	cem.-carpet 26.7	switches 27.6	suspended 7.5	door lock 39.2	handles 22.7	faucet 24.5
	painting 33.6	cem.-wood 39.0	wiring 31.2	gyp. boards 92.5	interior door 44.8	cabinets 43.2	rough plum. 31.4

Type of modification, by percentage (%)

Figure 4 Types of modifications in the basement

and Hardy. Soon after moving in, they set about using every bit of space to its maximum potential. After reclaiming a portion of the attic for storage adjacent to the master bedroom, they turned their attention to the basement, where they erected drywall partitions to create a family room and an additional storage area. In an ingenious use of space normally considered only for traffic circulation, the Laviolettes devoted .3m (one foot) of the 1.2m (four foot) width of the staircase to shelving for books. They had to build only one end of the support structure for the shelves since the other end of each shelf rested on the steps of the staircase.

Other new spaces in Grow Home basements included an additional bedroom, an additional bathroom, and a work place, some of which were also used as storage places. One resident created a living room in addition to a bedroom and bathroom in order to rent the basement to supplement her income.

We had assumed that the basement would become the grow area, and indeed it did: new rooms and uses were added shortly after the buyers moved

Figure 5 Typical new spaces in the basement. Clockwise from left: Washroom; office space; and laundry room

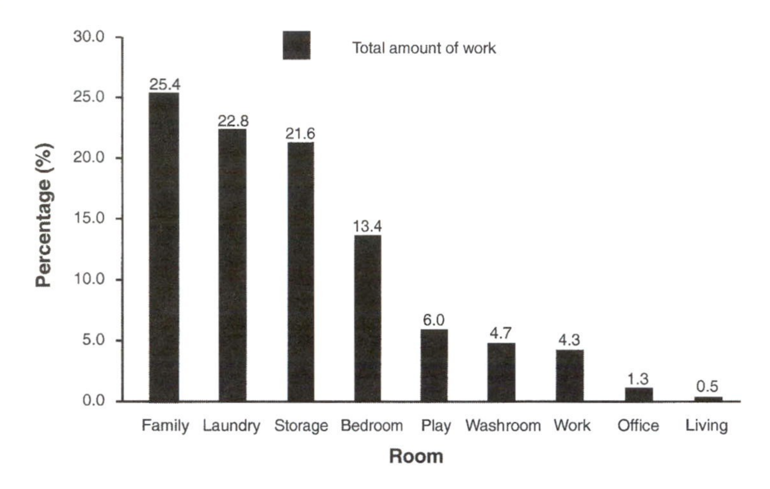

Figure 6 New spaces in the basement

in. We then wondered whether modifications would be carried out in other parts of the three-storey unit as well. What kind of work might that be? What would be the motivating factors for it? We decided to examine the main floor first and to study the living and dining rooms placed in close proximity one to the other. Other studies (Zeisel et al. 1981) refer to these areas in homes as "show off" spaces. We wanted to see if this would be the case with the Grow Home.

Only a tenth of the modifications performed in the house were made in the living room. We assumed that this low amount of work was due to the fact that most of the owners concentrated on creating a family room in the basement in order to leave the living room for formal activities only. Some examples of changes in this area included changing light fixtures to brighten and make the room look more attractive, carpeting the floor in order to provide a better floor finish, and changing the appearance of the walls with paint, wallpaper, woodwork, or mirrors. Although the living room and dining room share the longitudinal wall of the house, they are separated by a single step; some households made this separation more distinct by erecting small dividers or simply by applying a different finish to the shared wall. Another tenth of the modifications in the homes were made in the dining room. The primary change in this area was to the walls, including painting, wallpaper, woodwork, erection of a glass-block wall, and mirrors. Residents erected a low wall or wood balustrade to separate the living room from the dining room, changed the colour of the wall, made openings in the walls and removed the door leading to the basement to make the space look larger, and changed the appearance of the walls to make the space look more attractive (figures 7 and 8). Not having a formal dining room was one of the trade-offs that many owners made. They compensated for this loss with an attempt to formalize the space by differentiating the dining area from the living room.

The kitchen is a priority space in the lives of young families who tend to dine out less and not eat formal three-course dinners. We were therefore curious to find out what kind of changes were made in this key part of the home (figure 9). We discovered that the kitchen accounted for 10.4 per cent of all modifications. Some of the alterations involved changing cabinets and adding an island to obtain greater, more functional storage space, changing the floor finish to improve its quality, changing lighting fixtures to increase the amount of lighting in the room, removing the eating counter to gain additional

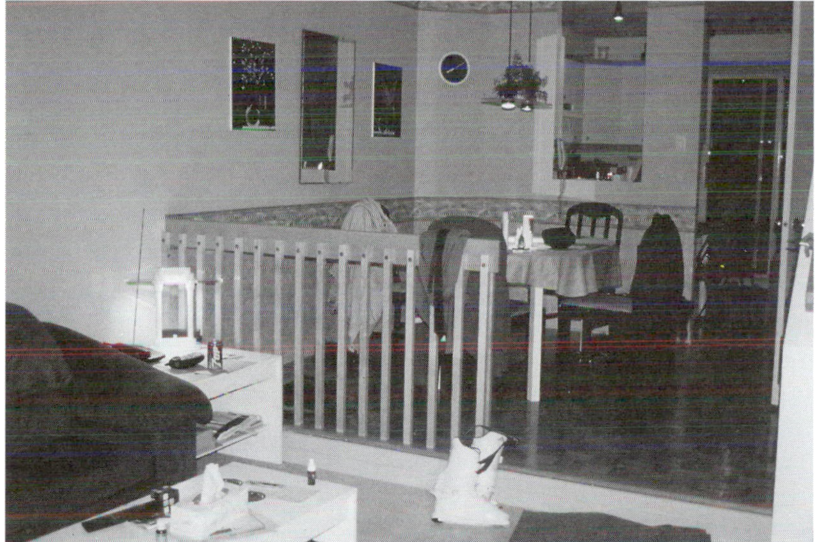

Figure 7 Separation of living room from dining room: homeowners erected different types of walls to separate the living room from the dining room, including a wood balustrade

Figure 8 Alterations made to walls: changing the openings, demolishing walls, and removing the door to the basement make the space appear more spacious and attractive, as well as improve illumination and ventilation.

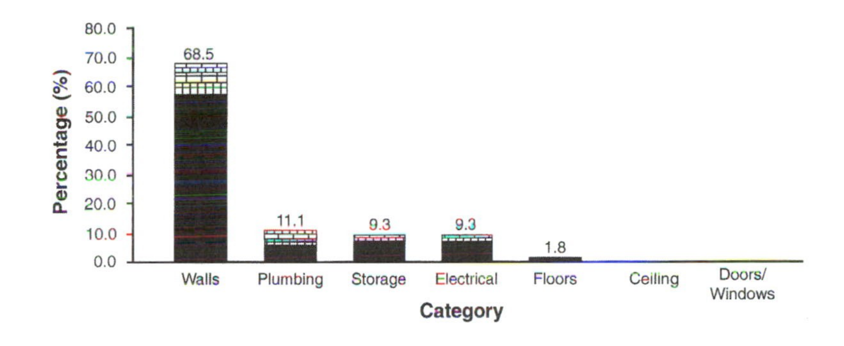

	tiles 1.2						
	mirror 1.2	rough plum. 17.6					
	woodwork 6.0	sink 17.6	island 8.3	wiring 8.3			
	wallpaper 7.2	dishwasher 17.6	handles 16.7	switches 25.0			
■	painting 84.4	faucet 47.1	cabinets 75.0	lighting 66.7	linoleum 100.0		

Type of modification, by percentage (%)

Figure 9 Types of modifications in the kitchen

space, and changing the surface of the walls for a better appearance (figure 10). Some residents remarked that they would have preferred to have a pantry or storage space rather than the eating counter which reduced the free space and functionality of the kitchen. The need for more storage capacity in the kitchen is the result of a change in contemporary food preparation patterns. Bulk purchases of ready-to-serve food in families with kids have increased the need for pantry storage. In many homes we found not only highly stocked cupboard spaces but a freezer in the basement.

Fewer resources had been invested overall in the main floor. The work was primarily superficial in nature and was intended to stamp a personal imprint on the space rather than satisfy a utilitarian objective as was the case in finishing the basement (with the exception of the kitchen, to a lesser extent). We next wanted to know what kind of work was performed on the upper, more private level where the bedrooms and full bathroom were located.

Since residents considered the master bedroom to be the space where they preferred to relax, they made it more comfortable through a variety of mod-

Figure 10 Changes in the kitchen: this household replaced the original kitchen cabinets and added an island to augment the storage and working space in the kitchen.

ifications. Work on the walls included painting, wallpaper, woodwork, and mirrors. Closets were converted and provided with light to make them more practical. Lighting fixtures were replaced in order brighten the room and to improve its appearance. The second bedroom was originally intended for children. Families with children were, in fact, the main group who made modifications here. A tenth of all changes made by residents in the entire house were performed in this room. The primary type of work involved walls, followed by storage, electrical features, floor, and plumbing features. Most of the changes to walls and electrical features were made to improve the appearance of the room, while others were made to obtain greater functionality. For example, one of the households converted the closet in the second bedroom occupied by their only child by installing a washer and dryer because they considered it too difficult to go down to the basement to do the laundry. Significantly, twenty-four of the twenty-nine residents who used the second bedroom for purposes other than as a bedroom were households without children. In these cases, this bedroom was converted to other functions such as an office, exercise room, study, guest room, reading room, and even as a sitting room, where a fireplace was installed in one home, part of the wall and floor tiled to make it more functional.

The upper-floor bathroom (the main bathroom of the house) also accounted for a tenth of all modifications. The primary type of work here again involved walls. Changes in storage, plumbing features, electrical features, and the floor were also recorded (figure 11). These modifications were made to replace the floor finish (from linoleum to tile), to add tiles on the walls for a superior finish, to add cabinets for additional storage space, and to change the shape of a wall for the sake of improved appearance (figure 12).

After we had reviewed all of the questionnaires, read the interview notes, and looked at the site sketches, we were amazed. The residents of the Grow Home had accomplished so much in only three years.

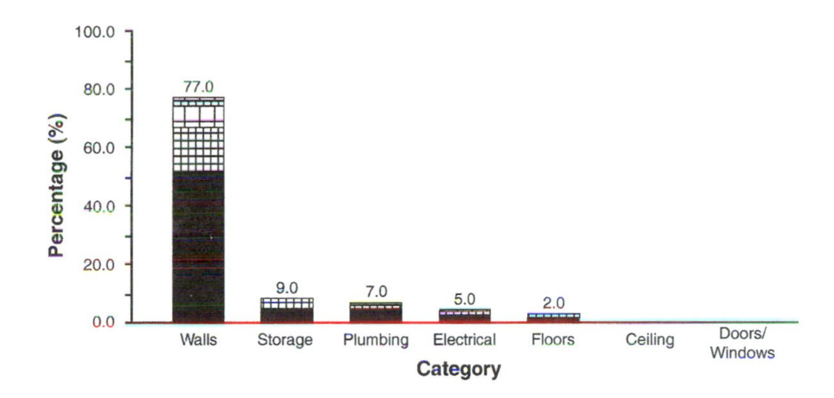

	tiles 1.0						
	woodwork 2.0						
	mirror 9.9		sink 14.3	wiring 14.3			
	wallpaper 19.8	handles 40.0	faucet 14.3	switches 28.6	lin. tiles 50.0		
	painting 67.3	cabinets 60.0	bathtub 71.4	lighting 57.1	linoleum 50.0		

Type of modification, by percentage (%)

Figure 11 Types of modifications in the upper-floor bathroom

There was hardly a home that had not been modified in some way. And far from simple weekend painting, the owners had done work that required not only developed technical skills but motivation and commitment to their new homes. The execution of the actual work was impressive. Almost three-quarters of all Grow Home residents indicated they did their own plumbing, electrical wiring, and wood floors. For an explanation of this trend, one need only visit a local renovation superstore on the weekend.

Many changes in the do-it-yourself field have taken place since the 1950s, primarily in the technology itself but also in communication to the public. Manufacturers of products and tools realized that they could reach more customers than just professional contractors. They could sell their products directly to the ultimate users, the homeowners themselves. To do this, they lowered the technical complexity of their items to the level of competence of most lay people. Iron and copper plumbing have now been replaced by flexible plastic piping that does not require soldering. Many products are sold as a comprehensive package of components cut to size and ready to install. When they are not avail-

Figure 12 Changes in the upper-floor bathroom Left: The shower wall was reshaped to make the room more attractive. Right: Tiles were applied to the wall.

able at the required size, the renovation store has a service centre which, for a fee, will cut almost any product to the client's specified dimensions. Materials have also been greatly simplified; ready-to-use tile cements, for example, have replaced complicated formulas for mixing compounds. Tools have also been made simple and affordable. For a relatively small investment a person can have a workshop with fairly sophisticated tools. Many people now buy their own power saws rather than borrow them from the only neighbour who owns one. The proliferation of knowledge about these tools and products has simplified their availability and use even further. Television shows, magazines, and seminars demonstrate how easy it is to renovate. Products come with clear labelling that provides a step-by-step description of the installation process of a sink or wooden floor. All Grow Home residents needed was the motivation to devote their weekends or after-dinner hours to work on their homes.

Many of the buyers saw the Grow Home as a first step, as a foundation in building their own equity. They knew that they would eventually attain their dream: a single-family home on a large lot. They also knew that in order to get there they would have to make their current home more sellable by increasing its resale value. They assumed that theirs would not be the only house for sale on the street and that a potential buyer would walk into several homes in the vicinity. Their home would have to be the most impressive, the shiniest. When we questioned the Grow Home owners about their reasons for renovating, 22.6 per cent answered that improving the house's appearance and its resale value was their first motive. This value was also a motive behind work by Grow Home residents on the outdoor areas of their homes. Three-fifths of all families with children put time and money into creating safe play areas in their backyards. Yet another reason to work on the backyard was to separate their newly acquired territory from that of their neighbours. After living in apartments and having a balcony as their only outdoor space, residents now had access to a yard. Some created elaborate gardens and decks where they would dine during the short Montreal summers.

There were also those who wanted the single-family home but were willing to wait longer. The motives for their renovation activities – in addition to improving the resale value – was to prepare the house for a second or even third life stage. More than half claimed that they did not see themselves moving out within the foreseeable future. When their young children were old enough to move into separate bedrooms, they would consider the basement. Highly pragmatic in their approach, these renovators envisioned their upcoming needs and acted to achieve their goals. The flexibility inherent in the structure of the Grow Home served them very well. The narrow width of the unit became an advantage, as they did not have to deal with space-limiting bearing partitions and had easy access to all the utilities.

The extended Lachance household took advantage of the Grow Home's flexibility to create a unique living environment for themselves. Denis Lachance and his wife bought a Grow Home unit in Cité Jardin Fonteneau in Montreal. Mrs Lachance's parents bought the adjacent unit. Denis and his wife had a home office in the basement and another room in the attic. When they had their third child and needed another room, they "borrowed" space from their in-laws by creating a connection between the two adjacent units: the back room in the in-laws' home became a bedroom for one of the children. The Lachances thus used the total volume of the two units to respond to their personal needs.

It was both instructive and exhilarating to observe the methods used by the residents to personalize their homes. As I toured these homes, the architect in me wanted to suggest other layouts, a different colour for a wall, an alternative tile pattern. I wondered why the bathroom tile was in a cascading pattern and not simply vertical, why the dividing wall between the shower stall and sink was stepped down. But I realized as I heard the residents discussing their homes with such immense pride that the space was theirs to do with as they wanted, not mine, and that they loved it.

7 THE GROW HOME OF THE FUTURE

The ability of Grow Home residents to make clear choices about spaces in their basements and then do the work themselves fascinated me. I was especially intrigued by the extent of their resourcefulness. They were able to articulate why a family room was constructed rather than a bedroom. They could also explain the present while plans were made for the future. Many even created lists of forthcoming tasks, matching events in the evolving family with building projects. Within an identical perimeter, each basement was different, reflecting a unique decision-making process. On the main and upper floors, the process was more one of personalization. I walked through the house units with homeowners, listening to their explanations, and I frequently heard them expressing their own design ideas. They wondered aloud about what they could have done as the architect: arranged the layout of the bathroom differently, added more storage space, reshaped the stairs in an alternative configuration. Some would pull out a pen to illustrate their plans with a quick sketch. I realized that buyers' knowledge about the design of their homes and their participation in the process cannot be marginalized. As I was trained as an architect to provide design services, I had some difficulty embracing the idea of buyer as designer. But I was certain about one aspect: more choice had to be given to the home-buyers before they moved in. Extending to them a single layout does no justice to their intelligence or their dreams. I recalled the process of conceiving the layout with Leo Marcotte and wondered why more choices were not offered in the first place. Why is there so little provision for change either before or after the occupants realize their needs are different from those they originally thought they had? There exists a paradox, I reasoned, between the dynamic nature of the twist and turn of our lives and the permanency, the rigidity embodied in the way homes are built.

Could Marcotte have offered his buyers more layouts? No, he could not. Selling eighty-seven units in two weeks in the middle of a recession was a welcome surprise for Marcotte. After his construction firm had sat idle for over a year, it felt great to get back into action. Yet he realized that the Grow Home project would be a tight challenge. His twenty years of experience had not prepared him for the delivery of eighty-seven units in three months. July occupancy was customary in Montreal, and construction could not begin until April. Marcotte himself had only three employees: a site superintendent, a bookkeeper, and a secretary. The construction work was subcontracted and Marcotte had friendly dealings with his subcontractors. (Before becoming a builder he had been a framer.) He realized an advantage he would have with his subs: they had been just as idle as himself during the recession and he would therefore have no problem engaging them and maybe even negotiating favourable discounts. The large building volume would provide an edge in the negotiations, given the economies of scale. The design of the Grow Home also worked to his advantage, with its simple rectangular structure and simple roof shape. Most of all, he was satisfied with his considered decision not to offer too many interior layouts and facade options at the time the units were sold. Since he knew that offering no options at all would probably hurt his chances of selling more units, he had offered a very limited range: two variations of interior layouts and two exterior designs. He was more generous with the offer of interior finishes which he knew from experience would be a good selling point.

The construction of the Grow Homes was swift and efficient. One after another, subcontractors came to the site to do their part. At times it looked like an assembly line with stationary products and moving men. The sequence of events was well orchestrated. Construction materials were delivered to the site to be ready for the tradesmen to install. Long identical prefabricated trusses were placed in front of each group of units, soon to become a roof. Identically sized windows were stored next to one of the houses. Large bundles of pink insulation were kept at one corner of the site and delivered to each house as needed.

Marcotte's building site followed the conventional stick-built construction process, one that has not changed much since the boom years following the Second World War. Builders like Levitt in the East and Eichler in the West recognized that efficiency of production was the key to coping with demand and to lowering cost. The development of modular prefabricated products was another evolutionary step that helped with efficient production. The assembly of many construction products can be made simple when they fit well with one another. Studs are spaced 405mm (16 inches) apart to accommodate batt insulation of the same width. The specific size of a sheet of drywall is designed and made to fit onto the centre line of a stud. The construction process has evolved to the point where products from different manufacturers can be assembled into a seamless unit. Modern builders have invented an exceptionally efficient method of on-site building where the assembly of products carries on, rain or shine.

It should therefore come as no surprise that despite the huge demand, the factory-built home has never taken off. Prefabrication has been attempted on several occasions, with only marginal success. Two particular cases are instructive, one from the first part of the twentieth century and the other from the middle. Several companies in the early 1900s offered prefabricated houses in a number of styles, the most successful of which was Sears, Roebuck and Company. Between 1908 and 1940, Sears sold houses through their mail-order catalogue and sales offices to nearly 100,000 clients. The company maintained its own mills and factories, and shipped materials for houses by rail to customers across the country, mostly in the Northeast and Midwest. The two principal lines of houses, the Standard Bilt and the better quality Honor Bilt, were precut in the factory and sent to the site as a kit of parts. Designs were chosen to reflect tradition and popular taste, and they came in a huge variety of shapes and sizes. Financing decisions and short-sightedness rather than poor marketing strategies or a substandard product contributed to the eventual downfall of the Sears house. Their liberal financing policies (i.e., very generous mortgages) did not take into account the possibilities of economic recession or reduced housing need. Determined to achieve the highest sales volume possible, Sears failed to consider the risks of selling at any cost (Sears, Roebuck and Company 1991; Stevenson and Jandl 1986).

At about the same time that Sears ceased production, two architects from Germany began work in the United States on a factory-made house later known as the Packaged House. Walter Gropius and Konrad Wachsmann developed their idea in 1941, including in the design concept the capacity for flexibility and variability. They received funding through private sources and government loans and guarantees from the National Housing Administration. They acquired a large war surplus factory which could produce thirty thousand houses per year, yet a very small number of houses was produced and sold. By the 1950s the venture had collapsed, again not for technical or architectural reasons but of causes related to marketing and to research and development. The competition with conventional builders of housing was too great (the public tended to choose the most affordable options), and the designers of the prefabricated housing wasted time and resources in moving the initial concept to the final production stages. Gropius and Wachsmann had used up half a million dollars before production began, leaving them without adequate financing. The unprofitable early years discouraged further investment and the business dissolved due to a lack of production capital (Herbert 1984).

There have been other attempts at prefabrication, but they never caught on with builders or the public. Is something inherently wrong with building homes in factories? Researchers in the field of prefabrication often maintain that the industrialization of housing holds many advantages over conventional construction methods (Robinson 1988; Kjeldsen 1988). The assembly of units, panels, or components under factory-controlled conditions, they argue, yields a higher quality product which generally results in more energy-efficient homes due to the optimum conditions under which the insulation is installed. The quick and efficient assembly on site reduces the effect of poor weather conditions, particularly in cold climates, as well as the potential for damage due to inadequate material storage and vandalism. Clean-up time and material costs are also reduced due to less wastage; construction management and trade coordination can be simplified; and the need for large teams of skilled on-site labour for multiple-unit construction is substantially lowered. While the potential for cost reduction is significant, particularly for standard designs and high production volumes, savings may be offset by delivery, installation, and inventory costs, as well as by higher fixed costs associated with keeping a plant under operation during the winter months when the demand is low and during years of reduced construction activity.

The use of prefabricated homes in Canada has been slow in gaining acceptance, although the use of prefabricated components such as wall sections is on the rise. On the other hand, factory-built housing is increasingly being used in countries such as Japan, the United States, and Sweden. In the United States, prefabricated housing accounts for 58 per cent of housing starts, with 15 per cent attributed to mobile homes and 37 per cent to panelized construction; in Sweden, prefabricated housing accounts for almost 90 per cent of housing starts (CMHI 1998; Cooke 1993). The reasons for the relatively small percentage of factory-built homes in Canada are varied and remain a question of debate. New priorities and, more significantly, new technologies have strengthened the belief that prefabrication will be attempted here once again.

So what does the future hold? If changes in the middle of the twentieth century gave birth to the North American suburb and the stick-built home, it seems that we are geared up for the next revolution. What is going to be the driving mind-set which responds to these trends? Flexibility and segmentation rather than homogeneity. It will be a unique approach to design, one that allows flexible arrangements of interior volumes. In a three-storey Grow Home structure, for example, a buyer would be able to purchase one, two, or three levels. A married daughter could live on two floors above her aging parents. In an adjacent unit, the lower floor could be a home office in a two-storey residence and the top floor rented to a single person. The structure would be sold by the level.

How will we buy Grow Homes? Gone will be the days of scouting for a home in a new development on Sunday afternoon. Using the Internet from the comfort of our own homes, we will be able to visit a future living room. The interior? From a digital catalogue, we will select components or prearranged packages as we design the bathroom that suits our lifestyle. We will be able to replace a bathtub with a multi-jet shower. A single person might wish to eliminate interior partitions, using furniture as dividers instead. No longer will choices be restricted to maple or oak kitchen cabinets. And to pay for it all? As we make up our minds, the computer will tally the cost, allowing us to fit the selections to our budget. Once the house is built, it will still be possible to revise the original layout decisions. If we want to give our unit a new look and change, say, the location of the dining room, it will not be a problem. The process will be explained in user manuals for the home. Buyers will be handed a CD-ROM along with the keys to the new home. The location of specially designed conduits in the floor and ceiling will be indicated in the plans on the disk. No more complicated than a step-by-step drawing that comes with a self-assembly furniture package, the disk will contain instructions, for example, on how to dismantle and reassemble a wall. We will depart from gypsum wallboard and wet joints as we enter the era of clip-on, pre-finished sheets. Homeowners could do the work themselves or engage a home service company, a new breed of business who help fit and change the house around.

Does this description of the future seem far-fetched? I do not believe so. Many of these technologies are already with us. Some, like the Internet, are in wide use. It will take time, however, until they are packaged and gain wide acceptance by the construction industry and the buying public.

Seeing the conventional stick-built construction on Marcotte's site, and listening to the occupants' comments on how they would have designed their own homes, I concluded that the small size, simple configuration, and efficient layout of the Grow Home would provide an opportunity to exploit the advantages of prefabricated building methods to their maximum potential. The construction already made use of standard material dimensions, thereby minimizing the cutting and fitting operations as well as material wastage. By using prefabricated roof trusses of standard size and slope, the roof construction was quick and efficient. Spanning front to back, these trusses eliminated the need for structural bearing partitions on the upper floor and made that space flexible. The built projects also made extensive use of prefabricated components: door frames, window units, kitchen cabinets, railings, exterior concrete stairs, and structural floor components were selected to simplify assembly and reduce construction time. The use of these components provided a starting point for the development of an industrialized version of the Grow Home whereby other prefabricated systems and subsystems (walls, floors, roofs) would form part of a complete system package.

There are three main prefabricated methods. *Modular* refers to the factory construction of sections (e.g., an entire house or part of one). The sections are sent to the site where they are hoisted into place by crane. *A kit of parts* is another method. Well-marked individual house components (e.g., two-by-fours, windows) are shipped to the site where they are assembled. The third method is to *panelize* a structure. Panels of different sizes, some with framing only (open panels) and others with insulation and windows (closed panels), are sent to the site and assembled according to plan. We chose a panelized system due to its similarity to conventional construction techniques. Since walls are normally made into panels on site, panelizing does not constitute a radical change from conventional practice and stands a greater chance of acceptance in an industry noted for its conservatism.

The adaptation of the Grow Home for industrialization was carried out in three stages. In the first, we selected a working model, based on four variations of the Grow Home concept that were built and sold in the Montreal area (figure 1). In the second stage, we established a set of criteria to guide the design process. Guidelines were drafted in three areas: architectural design, which responded to the occupant's expressed preferences and aspirations; modular standardiza-

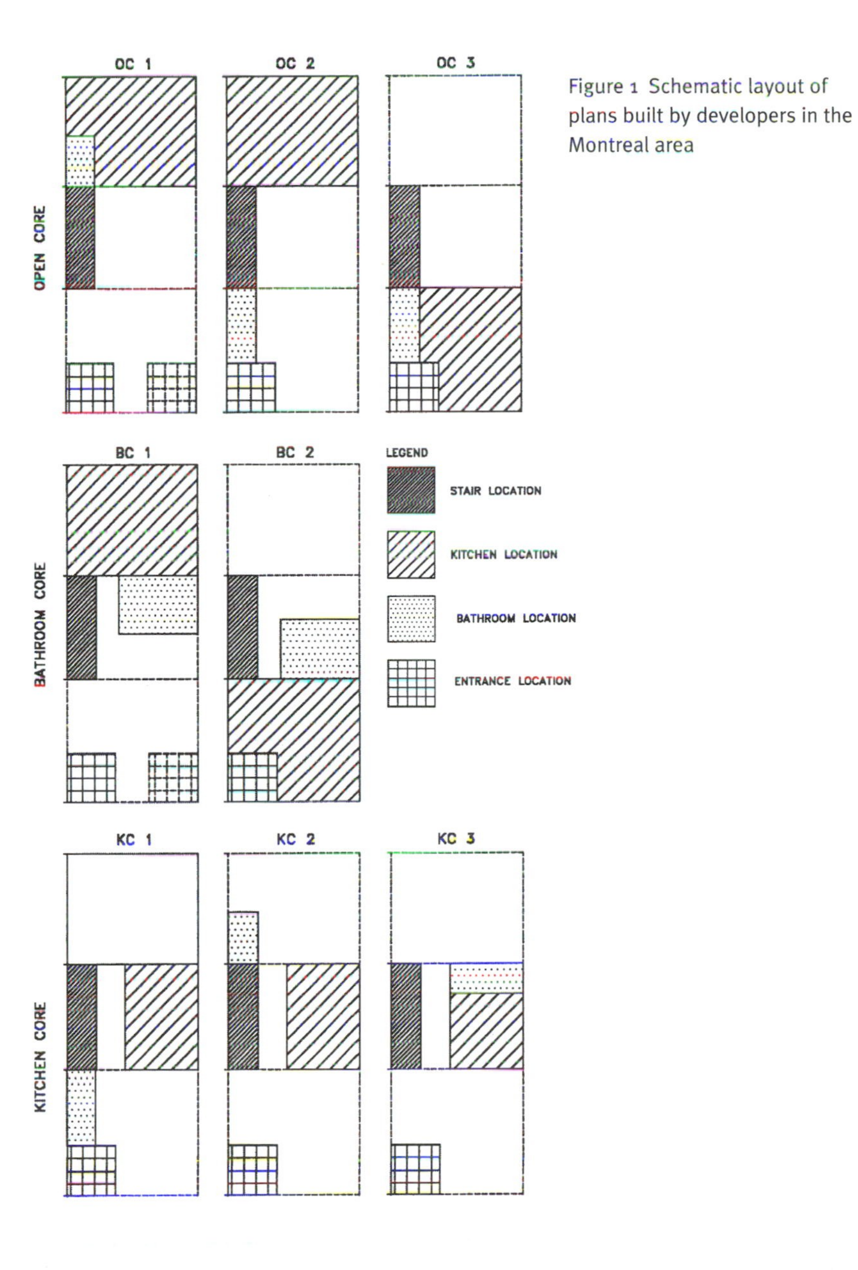

Figure 1 Schematic layout of plans built by developers in the Montreal area

tion, which addressed the prefabrication process itself; and technical factors, which were aimed at improving the construction quality. Finally, the design of the model units was optimized to conform to the architectural, modular, and technical design criteria.

The design process was aimed at providing sufficient flexibility for the builder, choice for buyers, and economies of scale for the manufacturer by generating a wide range of options for the dwelling and using a small number of simple, standard components. This process would enable mass prefabrication of components without the need to finalize the design. Modifications to the interior layout could then be made on site simply by adding or replacing components. The design process evolved from the inside out, starting with a general, basic analysis of the overall modular dimensions of the dwelling, followed by more specific configurations of the interior plan and ending with the exterior walls. This sequence was considered to be most suitable, since the flexibility and applicability of standard exterior prefabricated walls depend largely on the interior plan. We examined four aspects of the house: general dimensions, stair configuration and orientation, interior partitions, and exterior walls. Since the design process involved the manipulation of standard components within established modules, using a computer provided an efficient method of generating and testing alternatives.

Four basic arrangements for the interior spaces were generated for the ground floor based on the starting plans described above. These include units with a stair core (SC), an open core (OC), a bathroom core (BC) and a kitchen core (KC). For the basement and upper floors, the number of arrangements is limited by the types of rooms which are found at these levels. Only the first two options (SC, OC) apply to these plans (figure 2).

At the ground floor, the larger module (A) was generally intended to accommodate either a living space or a combined kitchen/dining area. The smaller modules (B, C) are sufficient for dining rooms, kitchens, or bathrooms. At the upper level, all of the modules can be used to accommodate either a bedroom, den, play area or bathroom. The design appears to be most efficient when the central module (C) is used for the bathroom while the larger of the remaining two (A, B) is designated as the master bedroom.

The manner in which the units were grouped, as well as their setting, affected the possible use of a space as well as the quality of light it was likely to have. Depending on the context in which the units are built, there may be a preference for a particular orientation. Builders may prefer to have bathroom cores back to back on adjoining units, buyers may want south-facing kitchens, and the addition of a window to the side wall of an end unit may require that a reversed floor plan be used. Figure 3 shows four alternatives that could be generated by reversing the units either front to back, left to right, or both. Reversing the unit front to back, for example, will dictate whether the larger module will face the front or back of the house. At the upper level this will determine where the master bedroom is located. At the ground level it makes it possible to enter the unit either through the kitchen or the living area.

In a narrow and simple unit such as the Grow Home, many aspects of the interior layout will depend on the stairs. The size, type, configuration, orienta-

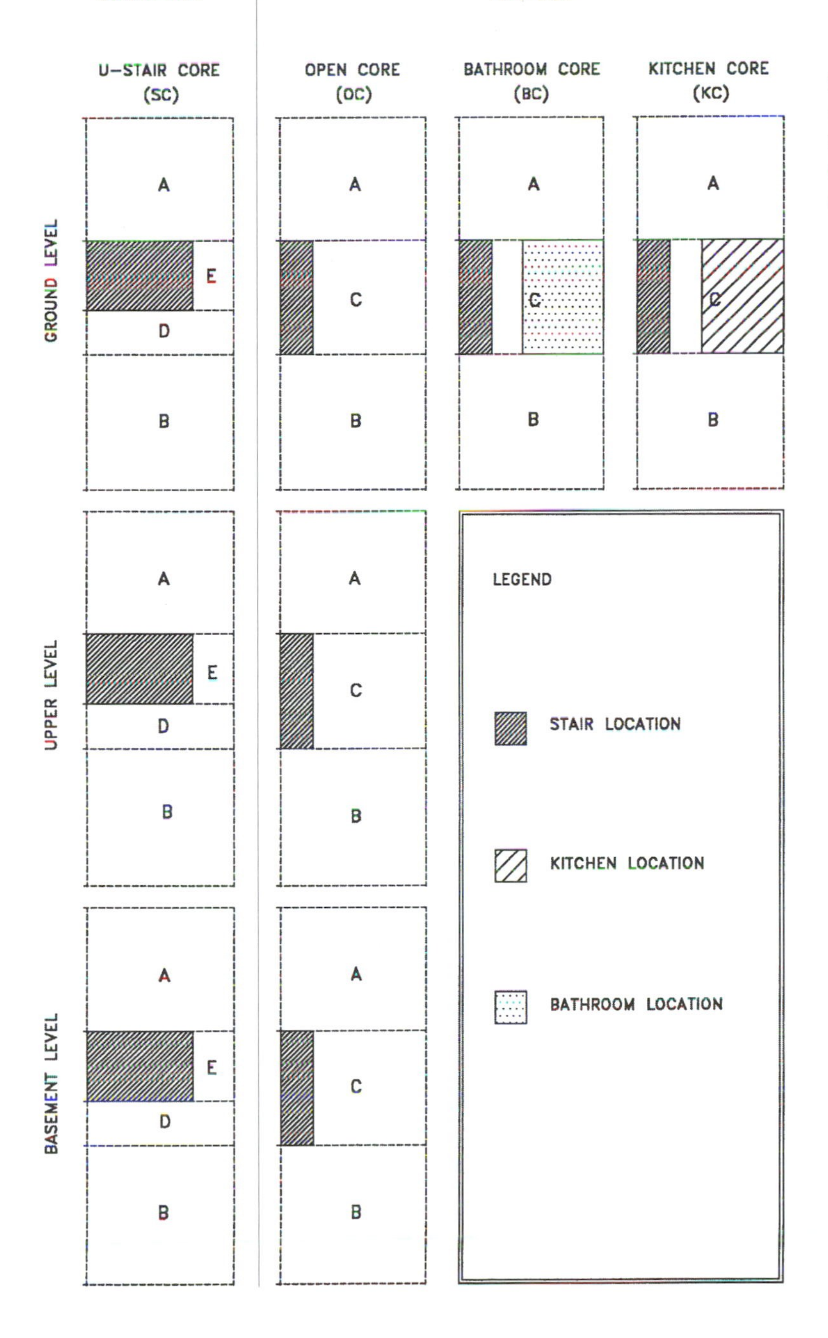

Figure 2 Segmentation of floor plans. (The letters on each plan represent spaces that can be created on each level.)

Figure 3 Reversed options

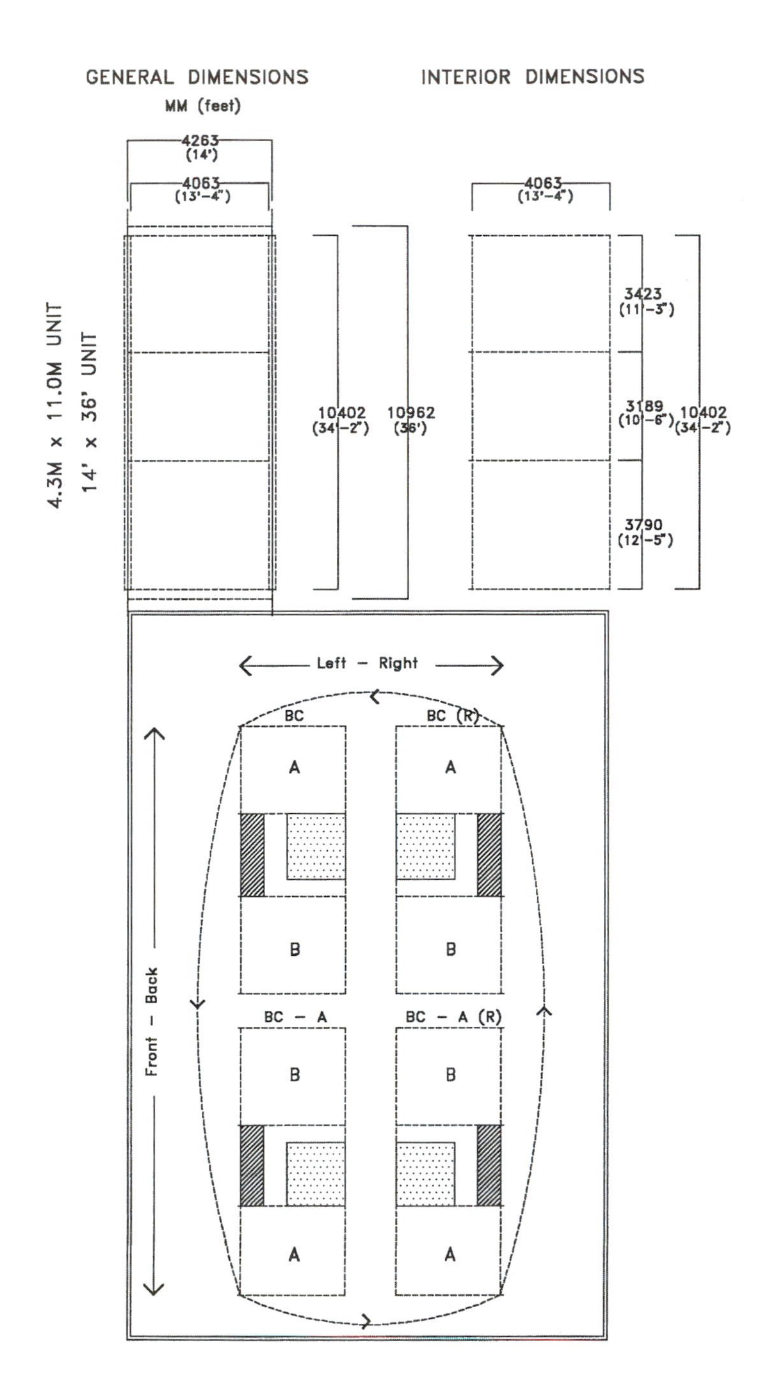

tion, and location of the stairs will affect everything from the size of the rooms to the general appearance and functional layout of the spaces. The ability of the builders to offer a variety of options and to make on-the-spot changes will therefore depend on the type of stairs chosen (figure 4). The construction of a staircase is usually more labour intensive than other framing tasks. Once built, it is not easily changed, particularly if the modification involves a different-sized opening in the floor. Prefabricating the stairs would enable several options to be offered for the same standard opening in the floor, and on-site changes to the layout could be made by either reversing or replacing the stair modules. In light of this potential for increased flexibility and standardization of parts, the stairs in the units were examined as an integral part of the prefabricated component system (in Canada, a standard riser/tread ratio of 7/11 is used, as specified in the National Building Code).

An adjustable modular steel staircase framing system developed by Stairframe, a Montreal-area manufacturer, is an example of the type of prefabricated stair kit that can be assembled to fit any set of measurements or design and can be installed quickly and easily. It is made of two lightweight steel components (a step support and a stringer) and a number of connectors. All the steps and risers can be installed with a twist of prefixed pegs. The staircase is put in after the wall studs and is screwed onto the studs, although it can easily be attached to concrete surfaces as well. The adjustable steel frame stays in place permanently while the temporary steel steps come unbolted to make way for the finished steps of the owner's choice. The advantages of the system are that it can accommodate any type of finish, allows for variation in width and step height, can be assembled and disassembled quickly to facilitate renovation, and costs 40 to 50 per cent less than traditional systems. Homeowners can select a design from a product sheet, or they can elect to create their own design.

Some of the implications of the stairs' configuration and orientation with respect to the interior layout of the units are illustrated in figure 5. Although many other options are possible, these diagrams serve to illustrate how the stair affects the potential for space usage and, ultimately, adaptability of the space. In general, the orientation of the stair (i.e., with a winder or as a straight run) will dictate whether or not the unit can accommodate a garage in the basement and affect several aspects of the layout.

After we dealt with and proposed an approach for the more macro issues of the Grow Home configuration, we moved on to the unit's interior design. Our idea was to develop a limited number of interior partitions that would permit alternative configurations as options. We envisioned a scenario whereby a builder would offer several options to the buyers. Alternatively, they would participate in the design of their own dwellings using the same elements. Communication with the fabrication plant would be simplified since only a limited number of walls would have to be produced. We proposed three standard dimensions – 610mm (2 feet), 915mm (3 feet) and 1220mm (4 feet) – based on our familiarity with the modularity used in house building and after much experimentation and study of the dimensions of the different rooms and uses in the conventionally built Grow Home. Although the assembly of small partitions in series may require

Figure 4 Options for stair configurations and prefabricated stair modules

SCHEMATIC STAIR CONFIGURATIONS

● departure point

→ arrival point

IMPLICATIONS

	FROM BASEMENT TO GROUND	FROM GROUND TO UPPER LEVEL	AT BASEMENT	AT GROUND	AT UPPER LEVEL
U-TYPE STAIR	1	1	with basement / without basement	with bathroom / without basement / various dispositions of rooms	two rooms / various bathroom and closet configurations
STRAIGHT STAIR with winder	2	2	no garage at front / one large room or two smaller rooms	various locations of bathroom / various dispositions of rooms	two rooms / various bathroom and closet configurations
STRAIGHT STAIR with winder	3	2	no garage at front / one large room or two smaller rooms	various locations of bathroom / landing with or without door / various dispositions of rooms	two rooms / various bathroom and closet configurations
STRAIGHT STAIR reversed	4	5	garage at front / one small room	various locations of bathroom / landing with or without door / various dispositions of rooms	two rooms / various bathroom and closet configurations

PREFABRICATED STAIR MODULES

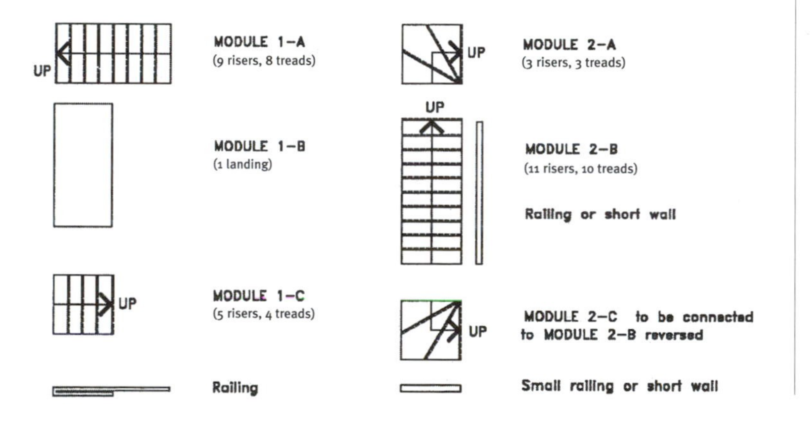

MODULE CONFIGURATIONS

1 = 1A + 1B + 1C

2 = 2B + 2A

3 = 2A rotated + 2B

4 = 2C rotated + 2B reversed

5 = 2B reversed + 2C

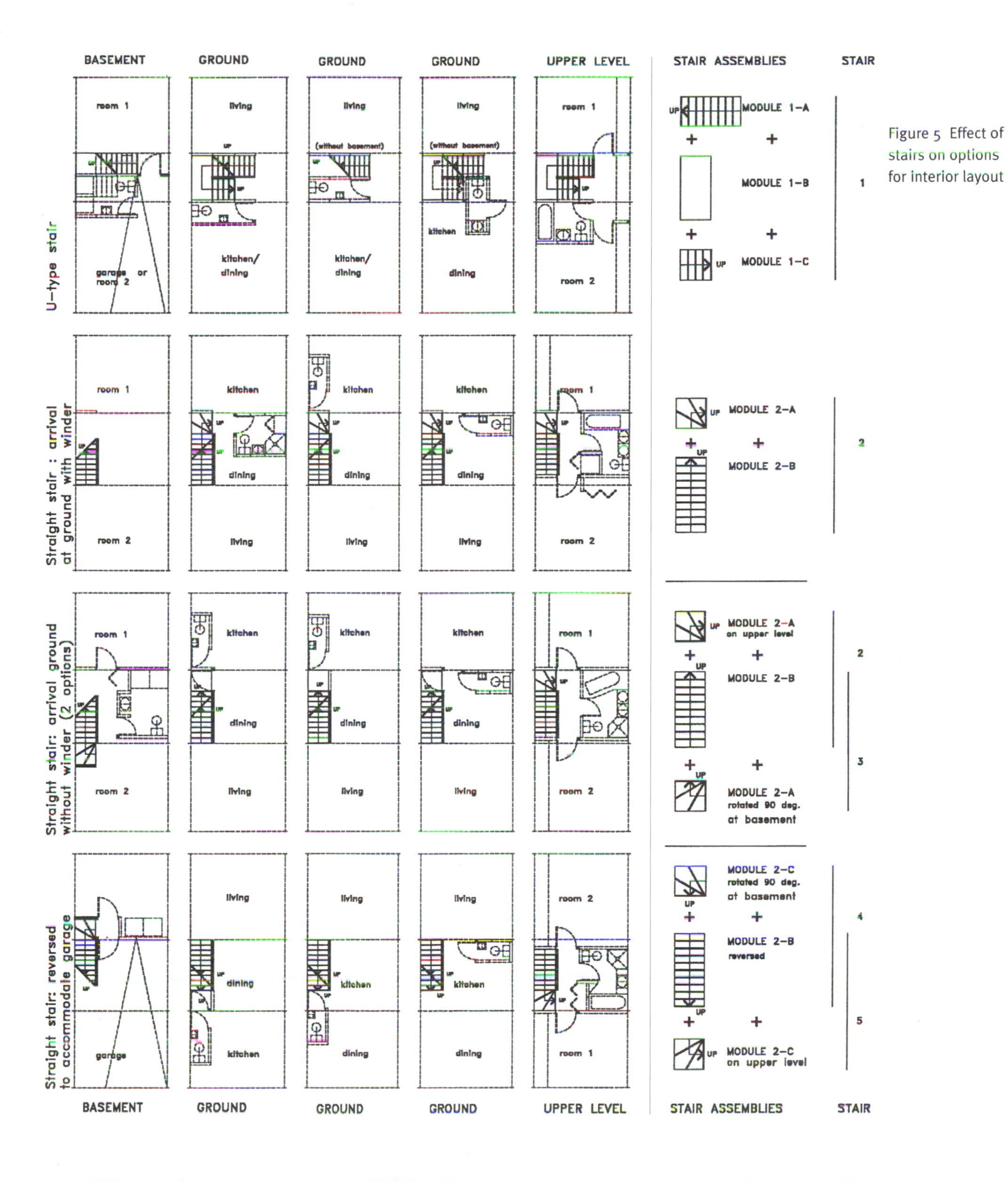

Figure 5 Effect of stairs on options for interior layout

Figure 6 Optional bathroom configurations using standard partitions

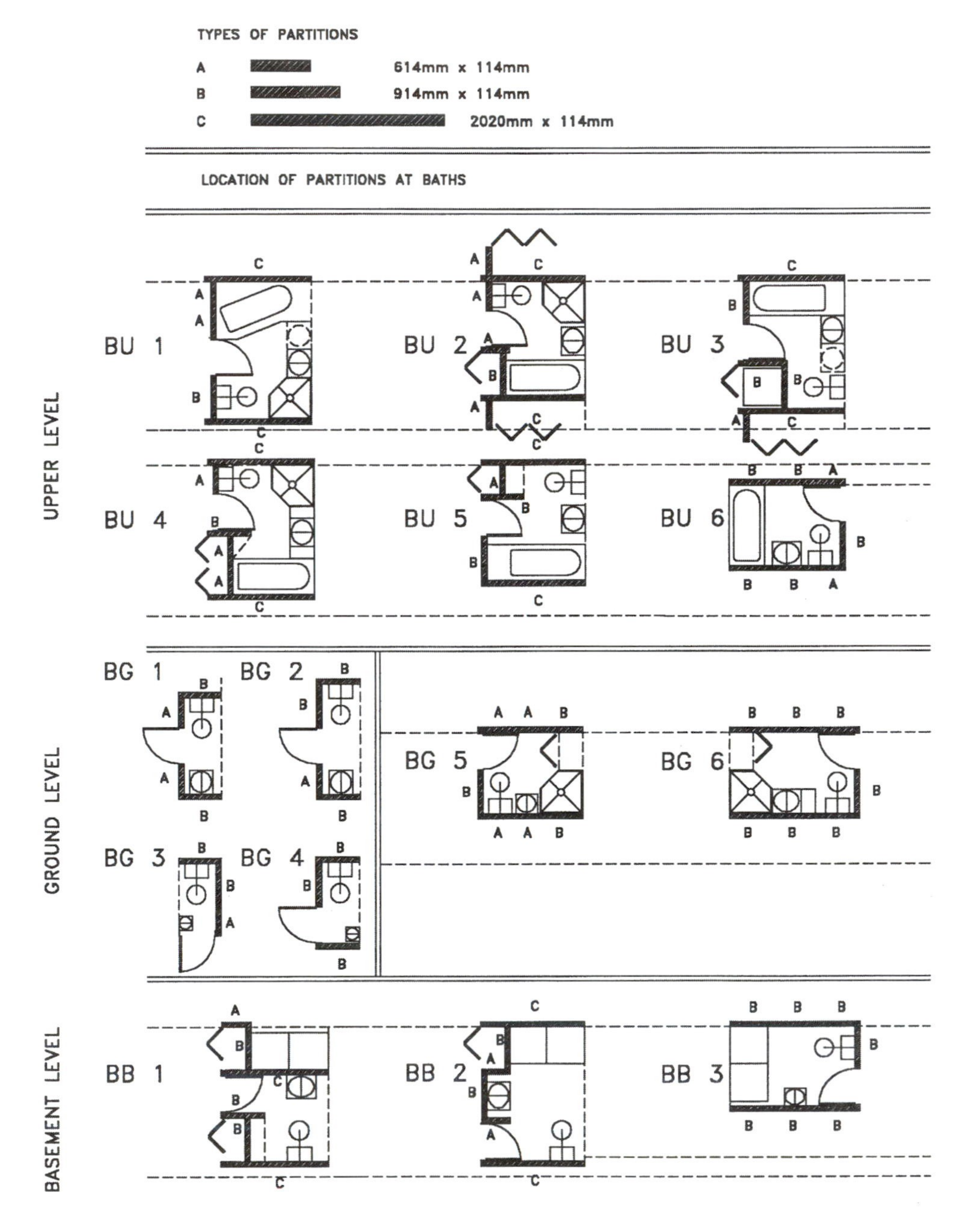

more framing members than in a continuous fashion, the increased flexibility, speed of assembly, and standardization may lead to economies of scale which might offset the added material costs. Figure 6 illustrates how fifteen different bathroom configurations could be built for the basement, ground, and upper levels using the same standard partitions. The alternatives include options for large tubs, double sinks, separate showers, linen closets and washer/dryer placement. Once the stairs, bathrooms, and entrances have been standardized, a variety of possibilities for the interior layout of the dwelling can be generated by treating these as modules in themselves. Examples of how these diagrams translate into floor plans are given in figure 7.

Personalization was important to Grow Home buyers. Facades are frequently part of residents' perception of their homes. It was unappealing to see in several projects that the same facade was repeated on all the units. We therefore took special care and paid close attention to the creation of a variety of facade designs. Will buyers design their own facades? It may happen eventually, but in the meantime we believed that greater choice should be offered. My conversations with builders have made it clear that a reasonable number of alternatives is not a drawback, particularly if they contribute to increased sales.

The design of standard exterior wall panels was consequently aimed at accommodating the range of options generated for the interior layout while addressing the cost-saving issues. The latter included the use of larger panels and simple, standard-sized openings located between the structural members of the wall system where possible (figure 8). Nine panel configurations were proposed in all: six for the front and back elevations and three for the side walls of the end units. Rowhouse versions of the home could be built with two to four panels, while semi-detached or end units would require three to six.

The biggest challenge in standardizing the exterior wall panels is keeping their number at a minimum while providing a pleasant, functional interior for each of the layouts which can be generated. There are two basic ways of reducing the number of standard panels. The first is to design them so that they could be rotated to suit the layouts of both the fronts and backs, left or right sides of the same or different units. The second approach is somewhat more restrictive, and deals with making the panels reversible. In this case, panels could be shifted from front to back or from side to side without the need to change their orientation. This would require that the panel be symmetrical in cross-section, a quality which is characteristic only of structural sandwich panels. Unsheathed structural panels could not be reversed, since they are either equipped with precut electrical chases or designed with an air space to accommodate electrical wiring, giving them a definite interior and exterior side. Figure 9 illustrates some of the plan options which could be assembled using various modules for entrance, bathroom and stair. The number of partitions and panels required for a particular design are shown in figure 10.

I had hoped to see a prefabricated Grow Home built in a project setting one day. I knew that in order to make it happen, I needed to find not simply a manufacturer but one who believes in the concept. In 1998 I got a call from Danny Cleary, a builder I knew from earlier times, who had built four hundred stick-built Grow Homes in a suburban town near Montreal.

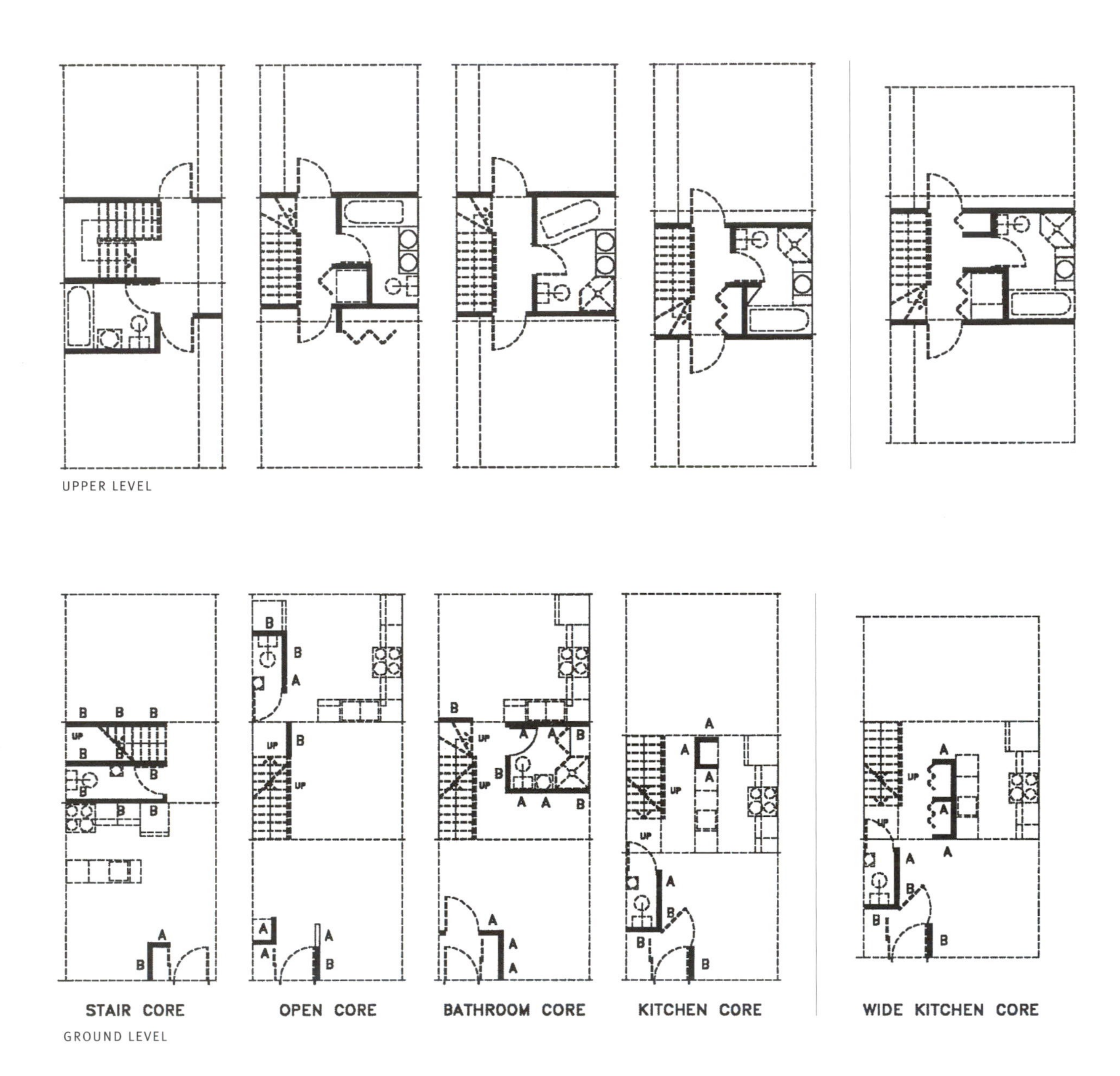

Figure 7 Location of partitions on floors

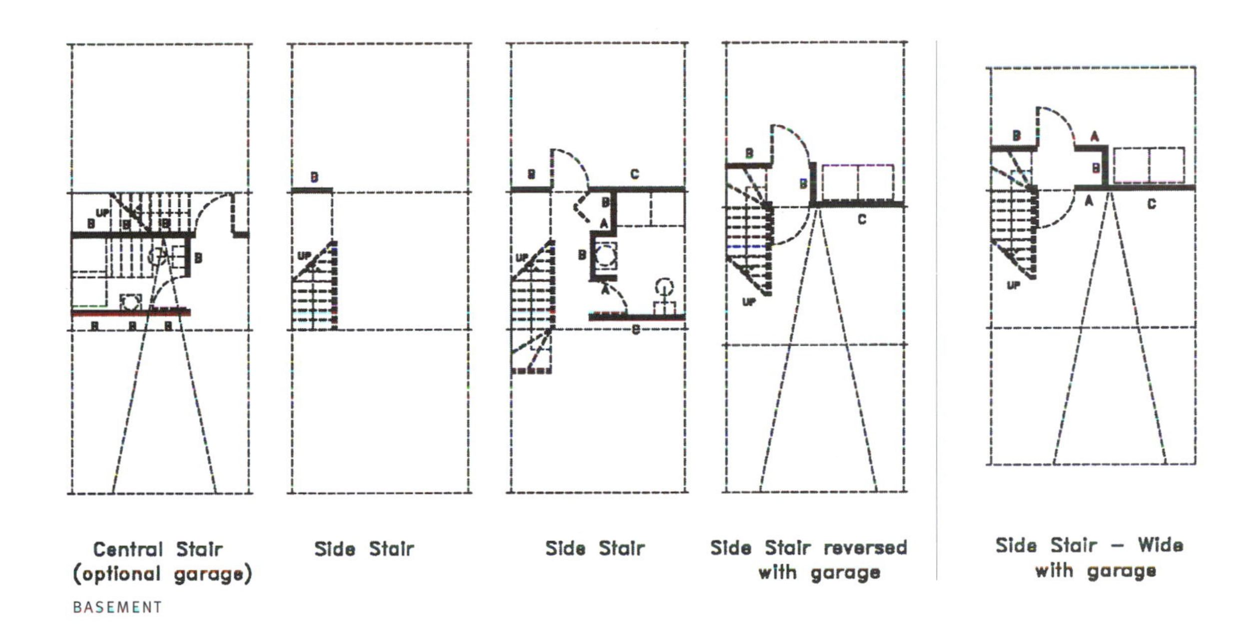

Figure 7 Location of partitions on floors

Figure 8 Standardization of openings in panels

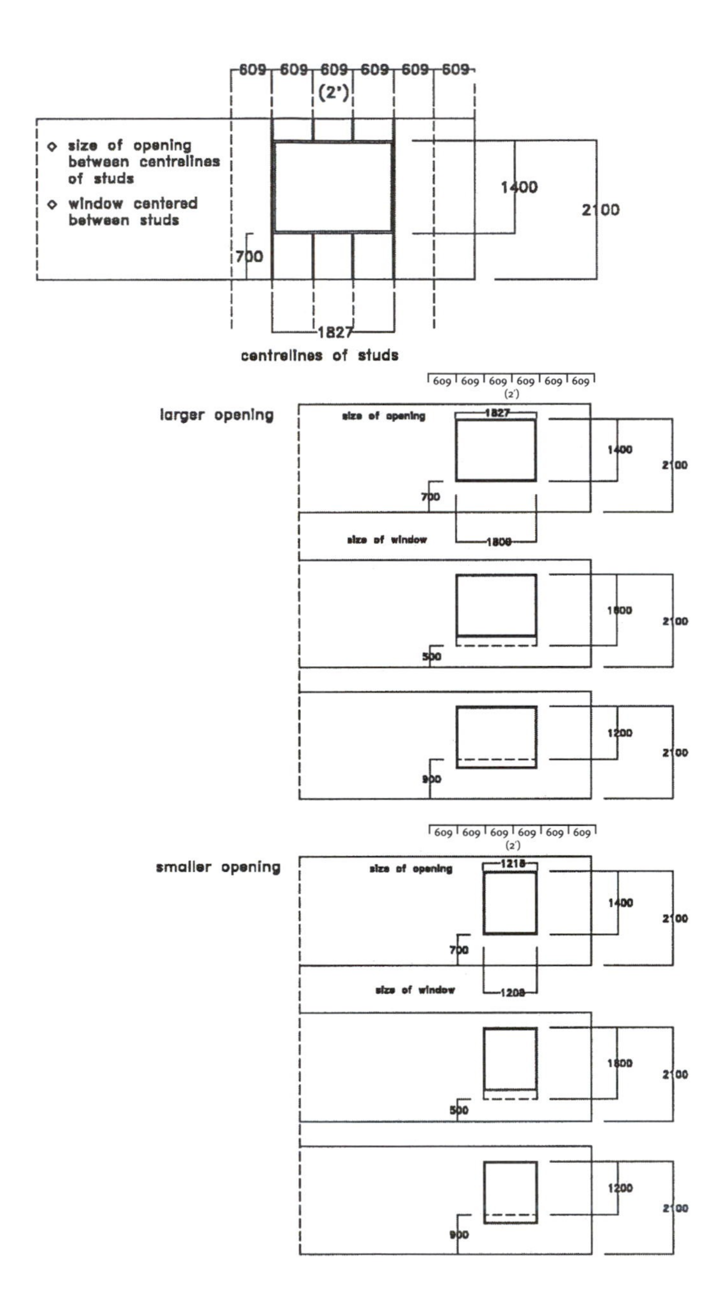

Figure 9 Sample ground-level arrangements using standard components

Figure 10 Components required for sample plan and elevation

Figure 11 Technology Building Systems plant in Ste Julie, Quebec, prefabricator of Grow Home units

"I want to prefabricate the Grow Home," he said. I asked him if he had a plant. "Not yet," he answered, "but I am going to find one." Technology Building Systems began operations in 1999 in Ste Julie, Quebec (figure 11). We developed the concept further and costed it. Danny Cleary now offers structural Grow Home packages for $8,000 (U.S.) per package. Designed for builders, it can be ordered and shipped anywhere in North America. Some time soon, I may see the day I had hoped for.

As you cross the u.s.–Mexico border from San Diego to Tijuana, the contrast between wealth and poverty is striking. Children shove boxes of chewing gum at the car windows, pleading with you to buy some. Young mothers from Chiapas in the south offer Mexican souvenirs as they point to the babies on their backs and the toddlers hanging onto their colourful traditional skirts. As you drive on, the California green is replaced by the brown of the valley ahead. Then you see them on the hillsides, hundreds, maybe thousands: brown dots as far as the eye can see. From a distance, it is difficult to make out what they are. As you get closer, though, you see the shacks. You hope that they are abandoned, remnants of a past disaster, until you notice that there are people going in and out of these homes and ramshackle cars driving among them. As you come into the city, the chaos is overwhelming: vendors announcing their wares, disorderly signage everywhere, and honking vehicles. You realize that you have entered another world.

I came to Tijuana at the invitation of Manuel Rosen, director of the School of Architecture at Universidad Iberoamericana Noroeste. I had met him at a conference at my university where he heard about the Grow Home and suggested that I come to share my experiences with his students. Intrigued by the shacks on the hills, I asked him if we could drive over to see them. This was in April 1993; a few months earlier the region had been inundated by severe flooding, and the damage was still evident. The wreckage of many poorly built homes was lying in the ravines. We drove to the edge of one of the settlements and walked through it. It was my first encounter with extreme poverty, on an enormous scale. A washed-out bridge replaced by a log was crossed by a woman holding bags. Improvised wiring to nearby poles connected the homes to the main electric line. Recycled tires were everywhere, used to hold back eroding soil or serving as front stairs to a home. A large water truck drove around slowly, stopping in front of each house. Women came out of their homes to fill up plastic containers, pay the driver, and haul the water into their homes. The dwellings had been constructed out of different materials. Some had walls made out of planks of wood covered with tar paper. Others had roofs made from corrugated metal sheets. The main doors and windows had evidently been recycled. Near one house, I saw a pile of old red bricks, waiting for reuse. Someone was building a new home out of concrete blocks near his wooden house. The whole area looked like a very sluggish yet active and giant building site.

A woman came out of one of the houses and greeted me. I climbed from the dirt road to her home which was clad with corrugated metal sheets painted green. The front of the structure served as a grocery store. Cans of food were neatly arranged on one wooden shelf, packages of toilet paper on another. Everything was spotless. The door to the back room was open, and I could see through to the living space inside. I made out a number of functions in a room no larger than 4.6 by 4.6m (15 by 15 feet). There was a small kitchen counter with a sink supplied with water from an overhead plastic pipe (no running water, no drainage). A large bed with a baby's crib beside it took up one corner. Another corner contained a table and four chairs. One section of the room was partitioned off with a curtain, behind which I saw a double bed – probably the parents' zone. I looked back at the smiling woman. This was her home. The same comment that I heard from many visitors of the Grow Home prototype now came into my mind: *This place is small*. The context was completely different from my own, economically as well as culturally; nevertheless, despite whatever background issues might govern the style and acceptable size of housing, I thought that there had to be a better way to house these people.

Walking down to the main gravel road, I asked Manuel Rosen about a large modern structure off in the distance. "That is a *maquiladora*," he said, "an assembly plant of a large multinational corporation. People come from all over Mexico in the hope of finding work at such a place. It is the only way to escape poverty in some villages." He went on to explain that when they arrive, the new workers rent or stay with relatives. They then squat on a lot and over time begin to construct their own homes. They do the building on evenings and weekends, either by themselves or with the help of family members. Informal electrical connections are arranged, and many years later, just before an election, a main water pipe might be laid. Rosen added, "These habitations are, in a way, quintessential Grow Homes."

I returned to Mexico in 1996, this time to Guadalajara where I was invited to lecture about affordable housing to local builders and architects as part of a home show. While there, I was offered a tour by a local private builder of a new, low-cost housing project. I happily accepted the invitation since I was eager to see how Mexico was coping with its huge housing deficit.

Despite a significant number of houses built every year in Mexico, the efforts to house all Mexicans are far from satisfactory (figure 1). The current housing deficit is in the range of seven million units,

Figure 1 A squatter settlement in Tijuana, Mexico

with an annual need of approximately 700,000 units (table 1). This demand is greatest in the twenty-five to fifty age group which constitutes 38 per cent of the population and is projected to increase to 48 per cent of the population by the year 2010. In comparison with Canada, where only 21 per cent of the population is under fourteen years of age, in Mexico the corresponding figure is 40 per cent, which will inevitably lead to a substantially greater demand for housing. The Mexican federal and state governments offer some aid through a variety of organizations and agencies which assist with building and financing; however, these programs are increasingly insufficient, a situation which is further exacerbated by the country's lack of a formal mortgage banking system. Mortgages are granted only to middle and upper income earners (those earning at least five times the minimum wage) at rates in excess of 50 per cent. The main sources of financing for a home are personal resources, government assistance, pension funds, and commercial banks for the middle and high end housing markets. Since the recession of the 1980s and a considerable devaluation of the peso in 1994, Mexico has experienced economic vitality. Middle and low-to-middle income groups are also gradually gaining economic strength, and a subsequent housing need is emerging as a prime area of activity (CMHC 1997).

To provide a sense of context within the Mexican

housing market, a residential building of 50m² (500 ft²) for a low-income family, including foundation, superstructure, exterior closure, roofing, interior construction, mechanical and electrical systems, architectural and engineering fees, permits and taxes, but without finishes, costs $22.89 per square foot or a total of $11,445. A similar structure of the same size, also intended for the low-income market but including toilet accessories and other basic finishes, an exterior brick wall, and waste removal, costs $31.98 per square foot or a total of $15,990. Finally, a structure of the same size, intended for the middle-income market, with all of the above features (many of higher quality) plus cabinetry in both the living spaces and kitchen and an exterior marquee, costs $47.87 per square foot or a total of $23,935 (Varela Alonso 1997).

When I arrived at the housing development for the tour with the builder, I saw a row of structures lining the street, all at different stages of construction (figure 2). The main building materials were cinder blocks, mortar, and concrete columns. The sloped roof was built of cement poured in place. Each unit measured approximately 60m² (600 ft²). It was a very hot day and there were only a few workers around, with no typical background sounds of hammering or sawing. We walked toward the end of the row, and I saw that the last unit was used as a model unit. Inside, a sales representative engaged in conversation with a young couple kept pointing to plans on the wall, probably explaining the layout to them. I toured the unit. The front kitchen consisted of a simple cement counter 1.8m (6 feet) long with a ceramic sink and water tap. I asked the builder where the cabinets were. He explained that the buyers would bring them when they moved in. The bathroom contained a sink hung on the wall, a toilet, and a bare shower stall. In addition to the kitchen and the bathroom, the house had two small rooms and a living area. There was also a small yard in the rear. The interior furnishings were of obvious poor construction quality: the metal-frame windows did not close well, nor did the door. With hardly any wood in the interior to soften up the place, the house had a barren feel. I was very surprised to hear that it took two years to build this home. In Canada, by comparison, to build a wood-frame house four times that size takes no more than three months. "Our building industry is not as efficient as yours," my host explained.

Year	Annual Need (Housing Units)
1996	671,318
1997	679,204
1998	687,375
1999	695,838
2000	704,608
2001	686,049
2002	691,990
2003	698,072
2004	704,288
2005	710,650

Table 1 Housing need in Mexico (CMHC 1997)

As we were leaving, I looked back at the young couple in the model unit. Now there were forms spread out on the table and the salesman was reading them carefully to the nodding couple. One of these units would soon be theirs, I thought. The house cost

Figure 2 New low-cost housing construction in Guadalajara, Mexico. Top: Units under construction using cinder blocks. Bottom: Finished units

Project	Price Range	Percentage of Population	Number of Families	Average Monthly Income ($ Cdn)
Low income	up to $6,590	68	11,000,000	220
Medium-low	$6,590 – 17,600	20	3,240,000	475
Medium	$17,600 – 33,000	7	1,133,000	815
Upper-medium	$33,000 – 82,400	3.5	566,000	1,660
High end	over $82,400	1.5	243,000	4,225

Table 2 Mexican housing values by market type (CMHC 1997)

$20,000: a fortune, a life's work for this couple. But the asking price for such poor workmanship did not make any sense at all. There had to be a better way to build than this, at a higher level of quality, in a shorter time, for a lower cost. If Mexico or any other developing country is ever to overcome its housing deficit, a different mind-set will have to be adopted. The odds against the people in these countries were, however, enormous. I knew that the challenge was not only technical but political and monetary.

On my return to Montreal I was approached by the organization that had arranged my trip to Guadalajara and my lecture there. They asked if I would be interested in designing a demonstration unit for the following year's exhibition at the same location. I recalled the squatters in Tijuana and the poorly built homes in Guadalajara and accepted. I was about to begin a sabbatical leave from the university, so the timing was convenient. My only condition was that the unit would have to be an affordable starter home for a young family. I wanted to demonstrate that significant savings in both time and money could be obtained using an appropriate design and the proper technology and process. I also acknowledged that despite my visits to the country, my familiarity with Mexican culture, habits, and concerns was weak. I therefore sought out a local architect with whom I could collaborate – Guadalupe Dipp, a Guadalajara architect and industrialist, who assisted me with the necessary background context and logistical on-site support.

I needed to establish a target cost when I began the design. I wanted to illustrate a design approach oriented toward the lower-middle-class market: employees with steady incomes, most likely dual-income families searching for access to affordable housing. Most purchasers would be funded and obtain a mortgage from their employers or a union as is the custom in Mexico. The cost of the unit would be between $13,500 and $20,000 which would specifically accommodate the chosen market profile. Table 2 demonstrates that a $20,000 unit falls within the purchasing capacity of the lower end of the medium-income market, a market defined by a monthly income of approximately $500 to $800. A $13,500 unit falls within the low-to-medium-income market defined by monthly incomes under $500. Taken together, the two market ranges encompass roughly one-quarter of the Mexican population (CMHC 1997). Based on average family income and construction cost per unit of area, I calculated that the area of the proposed home would be 30 to 50m² (300 to 500 ft²) – a small house, the size of one floor of the Grow Home. This would be the size of what the target market could afford.

The homes in Tijuana that I had seen were all basically add-on versions of the Grow Home. I would adopt a similar strategy in my own design: the start-up structure would provide the basic yet complete amenities for a young family, and more space could be added progressively. Building an expandable core house is common in the developing world. Countries which are unable to provide much-needed affordable housing for their large populations must economize on resources. The core provision may range from a site (plot of land) with very basic services such as a communal water tap, a main access road, and latrines to a site where, for a higher cost, a simple structure is also made available. Future residents are expected to finish the structure by themselves, much like the Montreal Grow Home. The housing provider is likely to be a housing authority (e.g., government or union) since the private sector concentrates primarily on the upper end of the market.

A design that incorporates distinct provisions for additions to the home must never compromise elements of the basic structure. For example, the relationship of the addition to the utilities and structure of the original unit is critical to the efficiency of the construction process. The original positioning of services is a major consideration with regard to a future addition. The effective organization of heating, plumbing, and electrical systems can significantly reduce the cost of an addition by enabling the builder or renovator of the new structure simply to link up with an existing system. Construction of an addition must also avoid costly damage to the original structure and ensure that the changes made are in the order of minor adjustments, almost cosmetic. A framework which supports these crucial considerations can be incorporated into the original structure by extending the foundation slab for a horizontal expansion, for example, or by fortifying walls to sustain a vertical expansion.

An addition must in no way compromise the habitability of the existing home. Close attention must be paid to maintaining the principal qualities of circulation, ventilation, natural light transmission, and acoustic insulation. In the interest of complying with these standards, the designer of the original unit must always consider future expansion and any associated constraints which could influence the configuration of the unit. The effect of these constraints can be buffered by efficient planning techniques: for example, by creating a configuration along the spine of a house instead of a central orientation which could impair access to the new addition or interfere with the existing layout by reducing light and ventilation.

Providing a very basic core was the thrust of the Tung Song Hong project in Bangkok, Thailand (figure 3). Designed by Chawalit Nitaya, the basic structure was constructed with hollow concrete walls and corrugated galvanized iron on top of wooden rafters (Nitaya 1981). The core also contained a toilet which

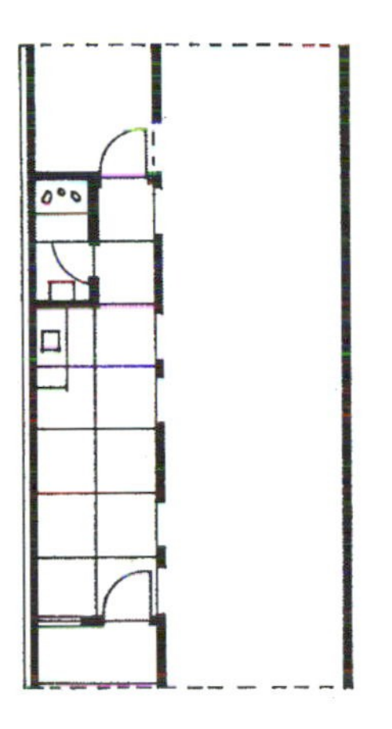

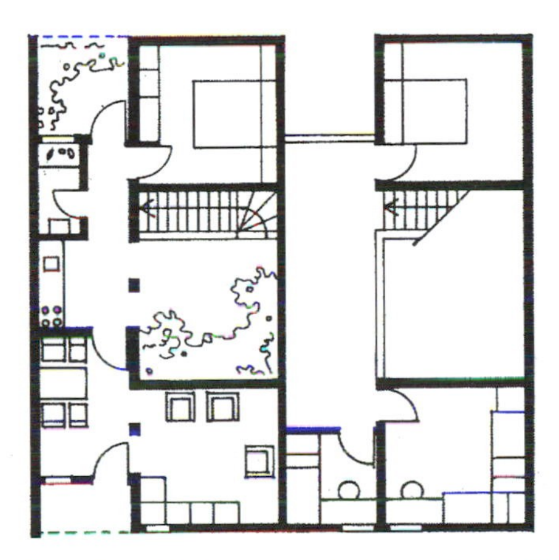

Expandable Minimum Ameriyah Dwelling, Egypt – Johan Buwalda, architect (Buwalda 1980)

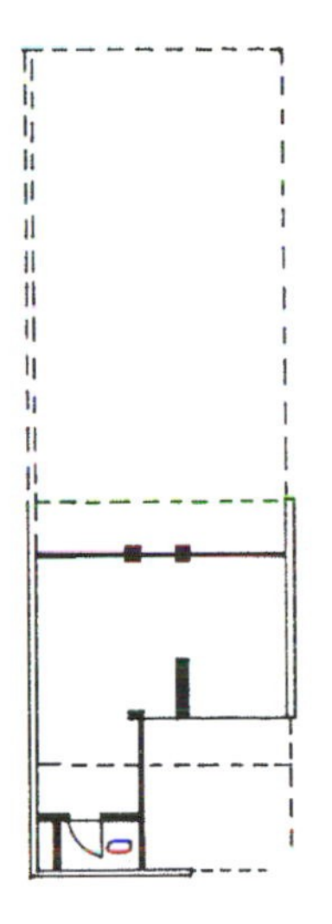

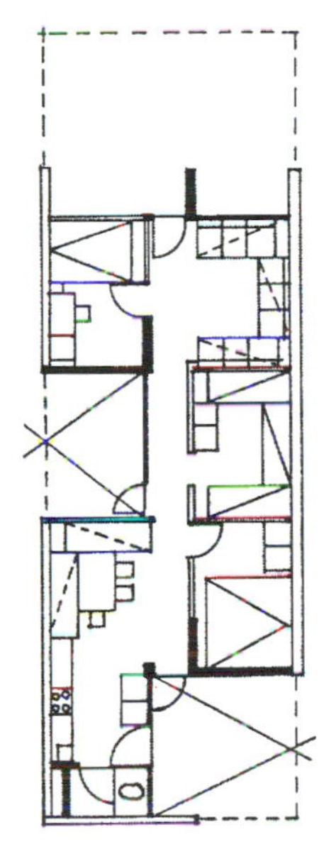

Tung Song Hong, Bangkok, Thailand – Chawalit Nitaya, architect (Nitaya 1981)

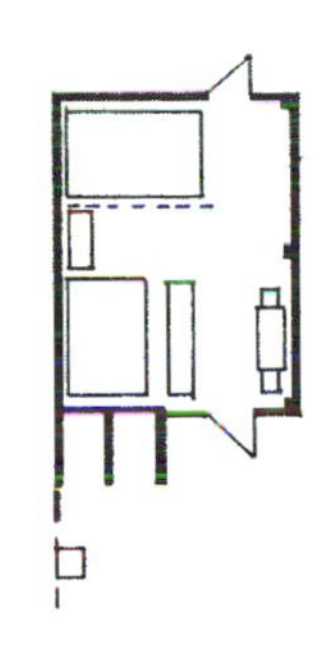

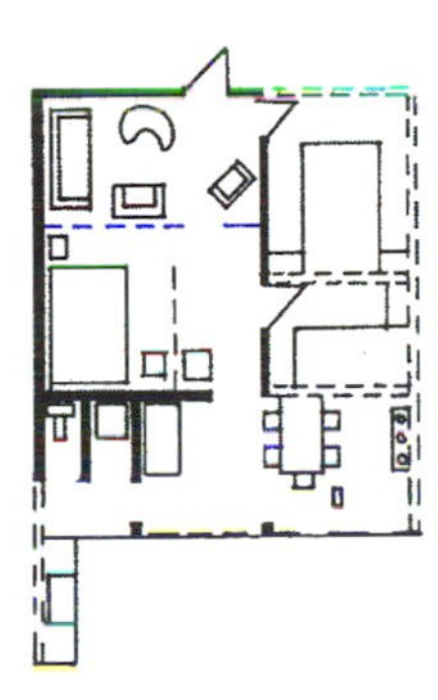

Tower Hill, Kingston, Jamaica – Shankland Cox Overseas, architects (Shankland Cox 1979)

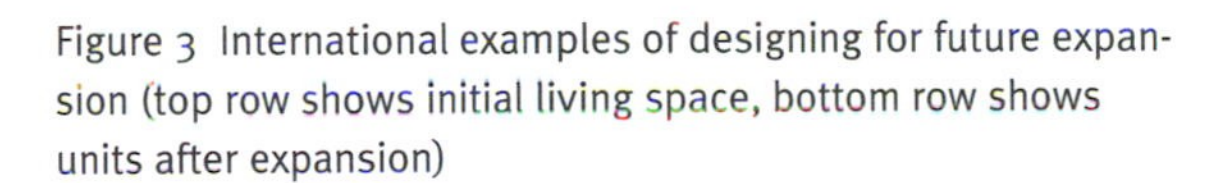

Figure 3 International examples of designing for future expansion (top row shows initial living space, bottom row shows units after expansion)

drained into a septic tank. Growth could either be horizontal or vertical. If expansion took place horizontally, the area of the house could increase to 80m² (800 ft²) from its original 30m² (300 ft²). A slightly more complete core was designed by Shankland Cox in the Tower Hill project in Kingston, Jamaica, where progressive building is part of the local housing culture (figure 3). The basic shelter was a room built of cement cinder blocks with a sanitary unit at the back that included a toilet and shower stall (Shankland Cox 1979). There were also a water tap and an entrance for electricity. The homes were built as semi-detached units, with a single water access and a single drainage access serving two units. The roof here (a challenge in such homes) was also constructed with corrugated metal sheets attached to wooden rafters. Expansion of the unit would be horizontal only, to achieve an ultimate size of 42m² (420 ft²) from its original 17m² (170 ft²). A further systematized approach was conceived for the Expandable Minimum Ameriyah Dwelling in New Ameriyah City in Egypt (figure 3). Designed by John Buwalda of the Netherlands on an urban grid of 36m (120 feet) and a building grid of 1.2m (4 feet), the project offered a range of module sizes to different income groups (Buwalda 1980). The long and narrow basic core measured 2.4 by 13m (8 by 44 feet) and contained a simple kitchen counter, a toilet, and a sink. The core structure was designed to expand sideways and upwards, leaving an open court in the middle.

When it came time to design the unit for Mexico, the principles underlying the Grow Home came in handy. As I had already calculated, the relationship between area and cost dictated the dimensions. To save on the cost of land and infrastructure, the units would be narrow-front and grouped as semi-detached or in rows. The design of the unit would have to anticipate the add-on process to fit in well with the domestic habits of lower-income Mexican dwellers. The price of the house could be lowered further with the introduction of do-it-yourself strategies in the interior. The residents would be able to improve the unit's appearance and quality as means became available. Both the internal and external design, of course, would have to reflect the values, culture, and habits of the region for which it was intended. A final consideration was meant to address the huge Mexican housing deficit: ensuring that the design approach included an efficient mode of production suited to the knowledge and capabilities of the local work force.

After establishing the key design principles, I moved on to what the initial basic stage of an expandable home should contain. Several concerns guided my thinking. The unit would have to be habitable and functional upon occupancy. It should contain all of the most elementary functions, even if in small portions only. It would also have to provide a measure of living comfort for its inhabitants. The place should be well lit, ventilated during the hot summer months, and warm during the chilly winter months. The initial unit would also have to meet its income targets and be affordable to a range of low-income users. Above all, it would have to be well constructed and above the current Mexican building standards.

Since local builders could very well be interested in offering more than one model to the public, I made a number of space and cost calculations based on data from Mexico. Four units were eventually proposed, with areas ranging from 33 to 50m^2 (330 to 500 ft^2) (figure 4). The designs for all of them were for narrow units where the functions had to be positioned with care. My collaborator Guadalupe Dipp and my Mexican students Miguel Rojano and Antonio Medina provided essential information on local domestic habits. The living, dining, and cooking areas were placed in the front of all the units. The four models differed in the number of bedrooms, the location of utilities, and the quality of the interior furnishings. Two alternative locations were proposed for the kitchen and bathroom: either back-to-back in the middle of the house or along the longitudinal wall. As we have seen in projects in other countries, when units are joined as semi-detached or as parts of a row, the same water and drainage lines can serve two units and result in savings of both time and money. Another important consideration – a lesson learned from the design of the Grow Home – was that sufficient storage space had to be allocated even in a small unit. And since it was common in Latin America to use private outdoor spaces for domestic activities, a utility sink was installed in the rear to be used for laundry or cooking.

The next step in the design process was to select one of the unit models for further development and for consideration as a built demonstration unit. We took an open-ended approach to the design to accommodate the practice of acquiring building components and installing them on an ongoing basis. A menu was therefore created to offer a selection of items ranging from storage and kitchen cabinets to bathroom fixtures and even large elements such as a pergola, parking shade, windows, and stairs. The

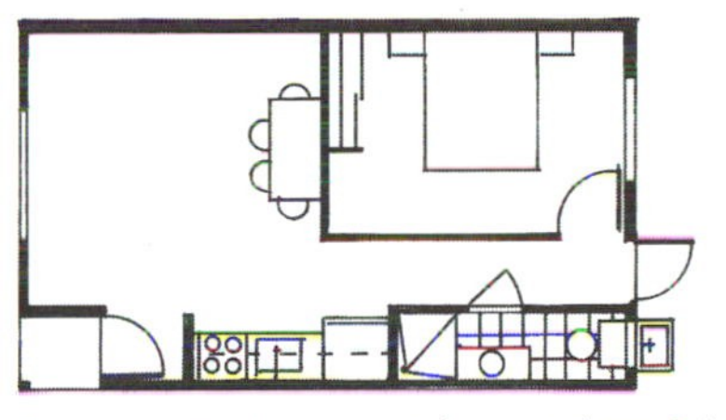

Small one-bedroom unit (33.1m^2 / 331 ft^2)

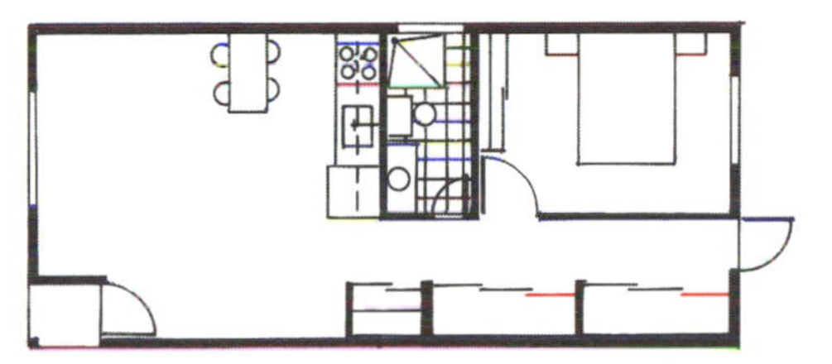

Large one-bedroom unit with back-to-back kitchen and bathroom (44.1m^2 / 441 ft^2)

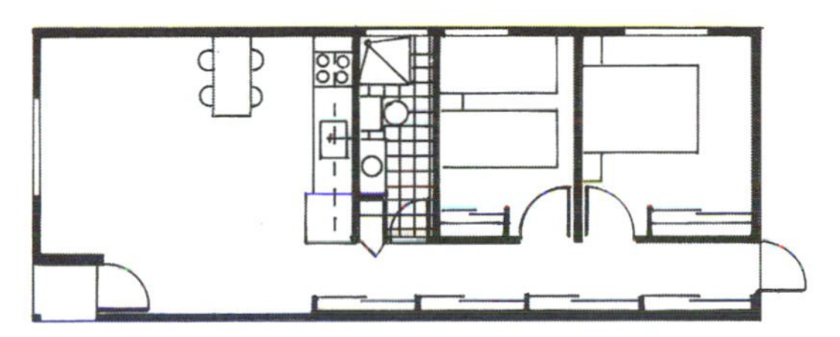

Two-bedroom unit with back-to-back kitchen and bathroom (49.6m^2 / 496 ft^2)

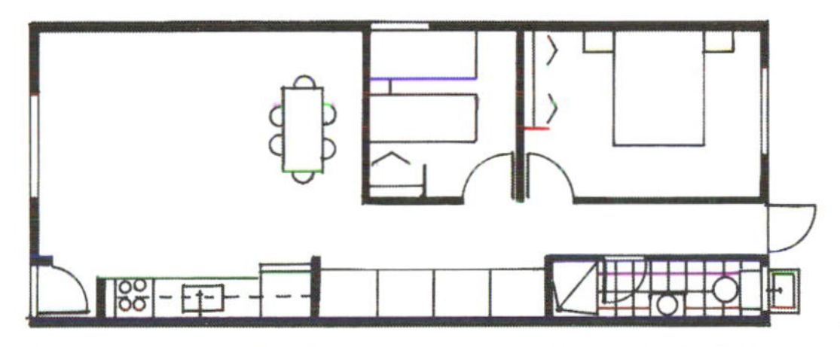

Two-bedroom unit with storage space between kitchen and bathroom to be used for future installation of stairs (49.6m^2 / 496 ft^2)

Figure 4 La Casa a la Carta basic floor plan: Four variations

Figure 5 Menu options: kitchen elements (sub-components in the two left columns and their combination into layouts on the right)

Kitchen Elements
Accessories:
Wall Rack:
Countertop:
Door Handle:
Sink:
Base Cabinets:
Drawer (C):
Corner Cabinet (D):
Pantry (E):
Layout Options:
A C B C
C A C B
A C B E
A C B E
E A C B D
C E A B D
B E A C C D
E B C A D
E C A C B D
Base Cabinets:
Cooktop Cabinet (A):
Sink Cabinet (B):
Wall Cabinets:
Cooktop / Refrigerator Cabinet:
Drawer Cabinet:
Sink Cabinet:

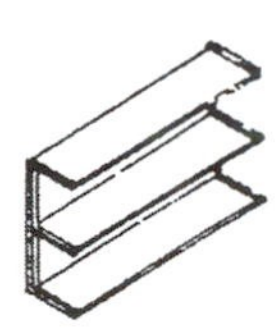

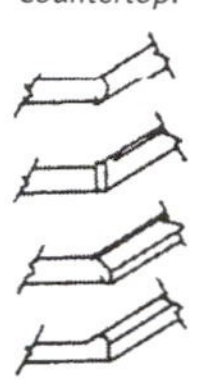

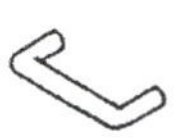

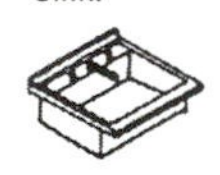

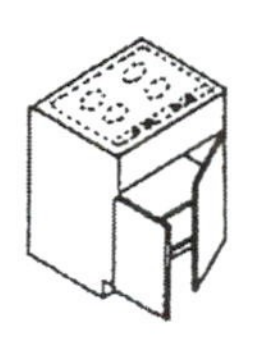

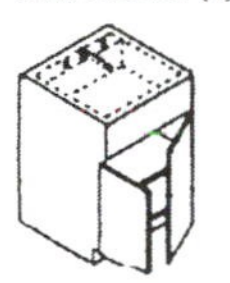

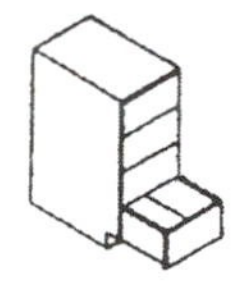

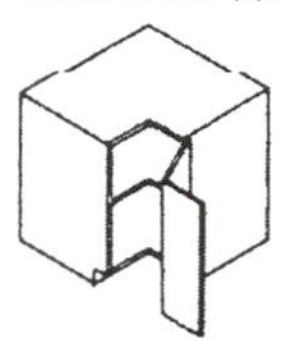

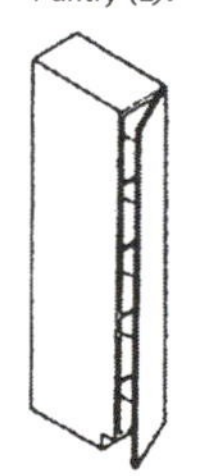

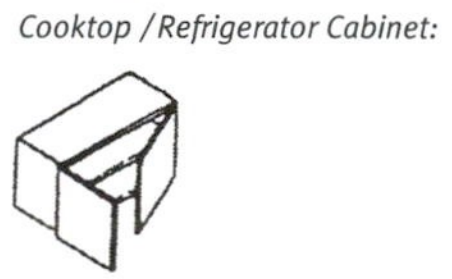

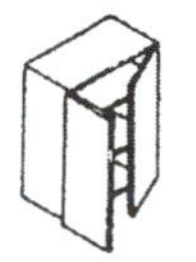

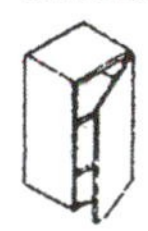

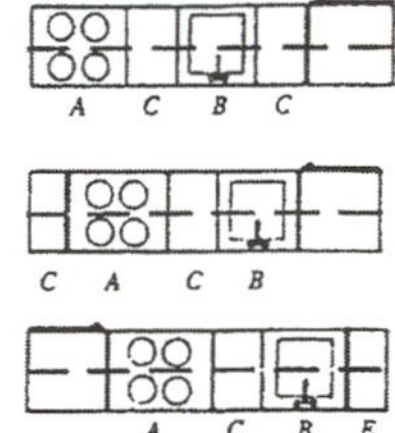

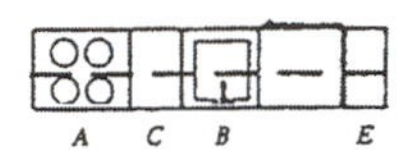

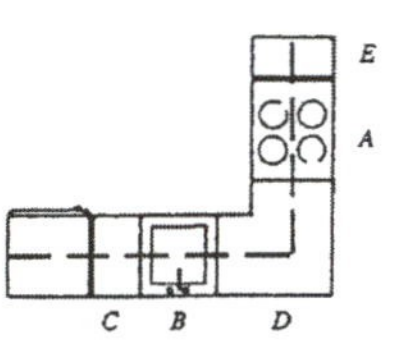

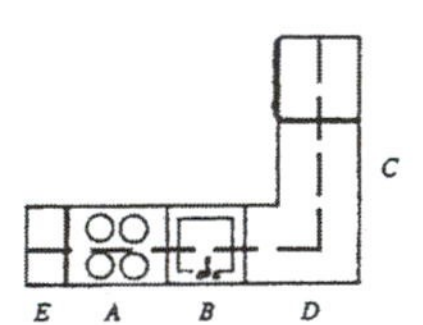

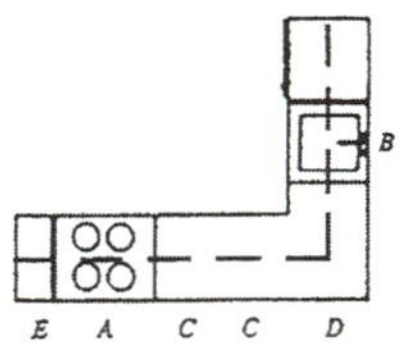

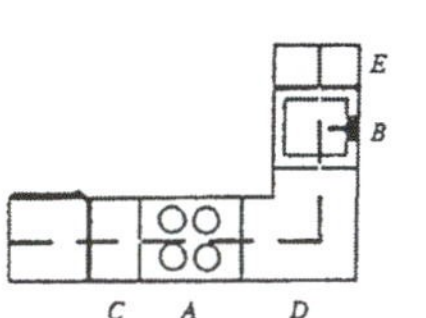

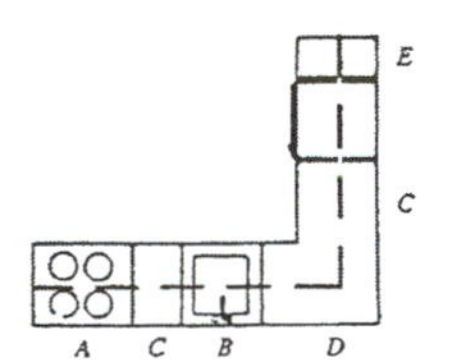

products were chosen following a rigorous research process to ensure that many were available in Mexico (Friedman, Horvat & Rojano 1997). The builder would be the first to make menu selections, limiting the range offered to buyers in a particular market and specific income bracket. We proposed that an actual menu of the options be put on display in a model unit to facilitate the buyers' choice (figure 5). This menu concept was so prominent that we named the house after it: La Casa a la Carta.

The accommodation of future expansion was implicit in the design of La Casa a la Carta. We anticipated potential modifications to features such as lighting, utilities, circulation, and construction (figure 6). The building process would likely involve self-help, requiring a flexible design with minimal complication to enable expansion by unskilled labour. Both horizontal and vertical expansion would be possible, and either could be built first. For horizontal growth, a courtyard would be needed to ensure adequate natural light and ventilation for the core structure as well as for the single large room or two small rooms that could be added in the rear. Alternatively, or as a second phase, vertical growth would begin with the dismantling of the storage component installed along the longitudinal wall to make way for stairs. The roof of the unit was flat and specially designed to serve as the floor of a future second storey. Once an opening in the roof was made and stairs built, three additional rooms could be built on the upper level as well as a bathroom built above the ground-floor bathroom to save on plumbing expenses. The second storey would add another 50m² (500 ft²) to the unit's overall living space. An alternative version of vertical expansion would locate the staircase within the external court area, either fully or partially covered. This positioning would allow the structure to be divided into two units: one entered from the front and one entered from the rear, particularly suitable for rental purposes. Expanding vertically and not horizontally gives the owner space for a backyard which could be used for a small business or for parking if connected to an alley.

An inevitable design challenge in such high-density development is to avoid repetitive, sterile, and monotonous environments. The facade design of La Casa incorporated the flexibility and individual identity that also governed the structure and plan. A composition strategy which included a variety of options in terms of appearance, style, fenestration and materials created a necessary balance between flexibility and individual unit character on the one hand and a measure of control in unifying the neighbourhood on the other. Providing a variety of facades and design options, grouping units in small numbers, and allocating appropriate parking arrangements and open spaces are fundamental elements of comfortable neighbourhood developments (figure 7).

When we moved on to the community design, we used the La Casa unit as a flexible building block. To be viable, this approach would have to deal satisfactorily with three connected issues. The first concerned the built environment and the appearance of a number of these units together: might they end up in a long, sterile row? The second issue involved circulation: how would people and cars move around and among the units? The third issue addressed the design of public and private open spaces: where would

Figure 6 The growing process

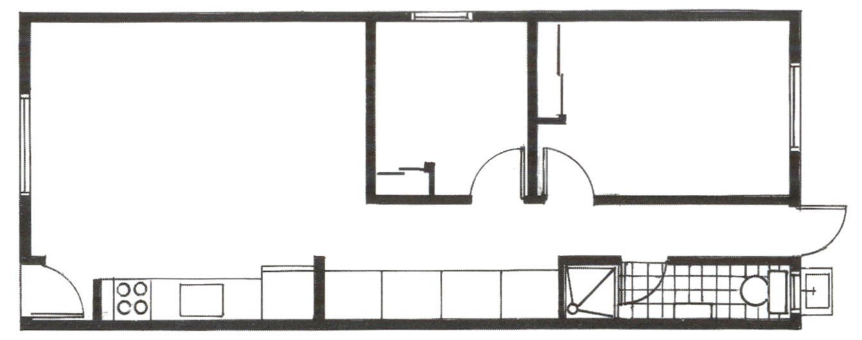

Initial stage ($49.6m^2$ / 496 ft^2)

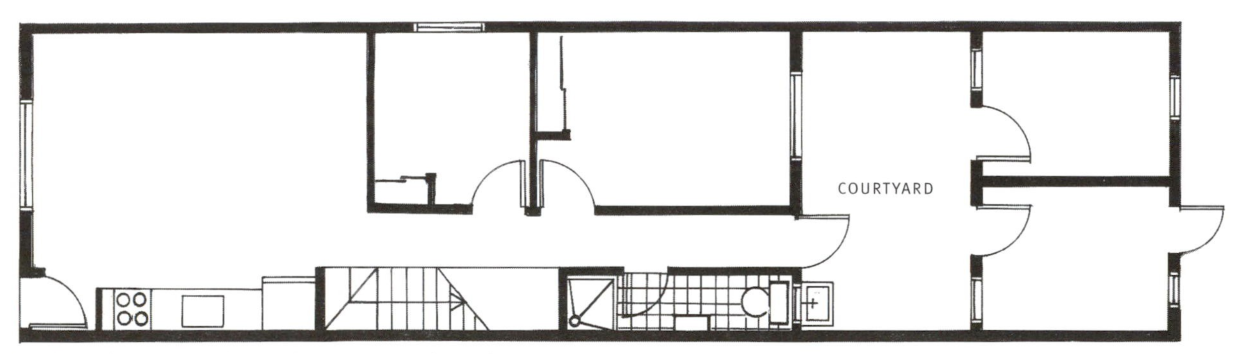

Horizontal expansion of an additional $12.4m^2$ (124 ft^2), with two rooms in the rear enclosing an open court

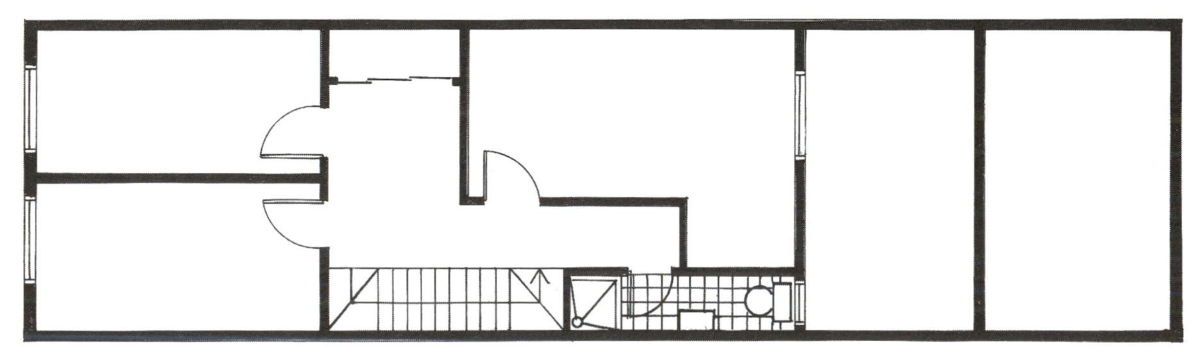

Second floor of a vertical expansion of an additional $49.6m^2$, where stairs were added on the ground floor to provide access to the new floor above

Figure 7 Facade options: three front elevations

Figure 8 Rowhouse units can be either staggered or aligned

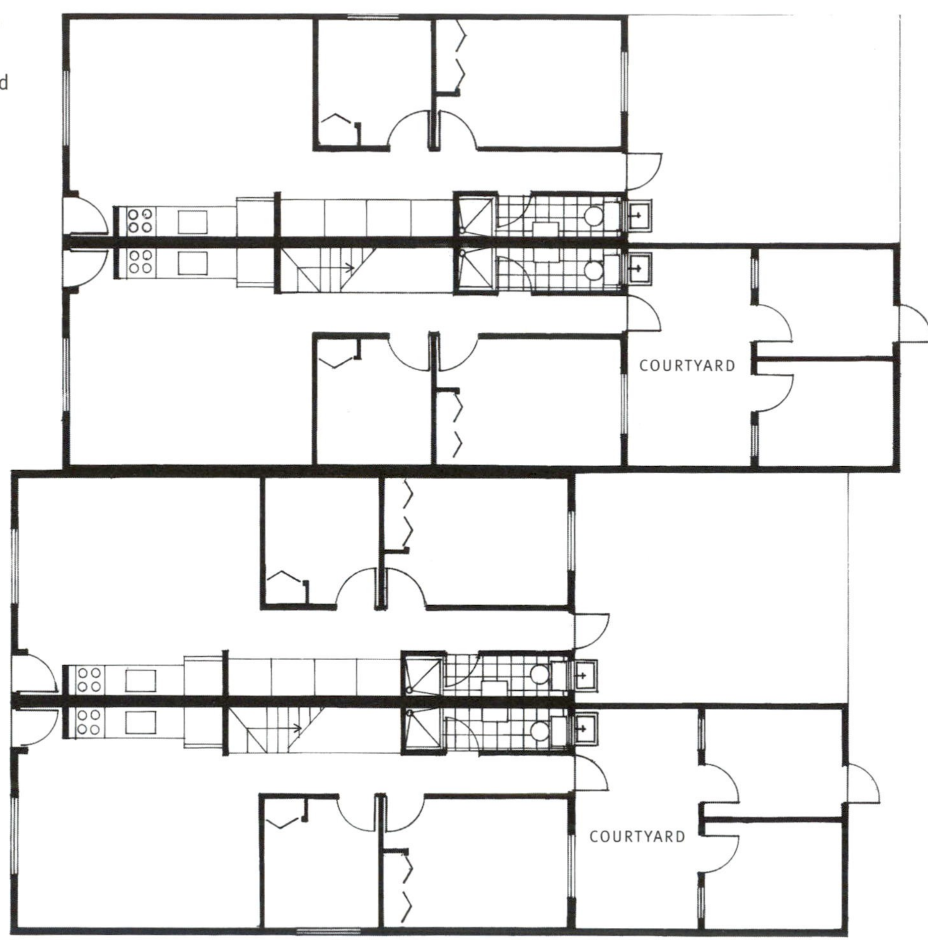

The wall separating bedrooms is not full height in order to augment illumination.

communal facilities and public parks be located?

The balancing of these three vital elements ensures the provision of pleasant and desirable housing as opposed to the type of neighbourhood that homeowners strive to avoid. In order to lower costs in higher-density communities of thirty-seven to sixty-two units per hectare (fifteen to twenty-five units per acre), housing authorities using prefabricated components in developing countries often repeat many exterior features. However, a fixed number of variable elements can be used in interesting ways to accentuate the identity of each unit. If a linear arrangement is proposed, grouping units in blocks of four to eight is preferable (figure 8). Street corners can be delineated by taller, two-storey homes and, when the code permits, commercial enterprises can also be placed on these corners to create communal meeting places. Clustering units around a large open space is another approach that can be used in high-density, low-rise developments where each cluster maintains its own character and identity with a common open space surrounded by private spaces located behind each unit. On a broader scale, a community planner may implement a design configuration that anticipates future modifications without compromising the overall visual harmony of the neighbourhood.

The number of cars per capita in Latin America is substantially lower than that of the vehicle-oriented societies of Canada and the United States. In Mexico, for those few people who own one, the car is frequently used as a source of income. For reasons of security, it was common for residents to park their cars as close as possible to their homes. Two adjacent homes could easily accommodate a single parking shelter between them, and if a back lane were available the car could be parked close at hand and its impact on the streetscape would be reduced.

Increased density often compromises the allocation of public open space. In Latin America, outdoor spaces are commonly used for domestic functions, and this area must also incorporate adequate area for future expansion. The lots must therefore have a minimum depth of 9.1m (30 feet) or, alternatively, a front yard must be provided. The front yard is an important feature of the rowhouse neighbourhood since it functions as the transition zone between the private and the public areas of the property and unites the home to the rest of the community.

Once the design was complete, we prepared for the construction of La Casa a la Carta. The open-ended approach to its conception was also reflected in its implementation. Some twenty Canadian manufacturers provided products that we determined to be suitable for the target buyers. In a real project, a housing authority or a private developer who undertakes large-scale projects would be the party who offered these products. To ensure compatibility and ease of application, we made certain that they would be suitable for all four options of the initial-stage unit. The first step in our selection process of these products was an examination of alternative structural systems.

In April 1997 I was introduced to Serge Maheux, the president of Archimède 2000, a Montreal prefabricated panel company. Maheux had developed expertise in lightweight sandwich panels and used them in several housing projects in Latin America. The core of the panels was light-gauge steel studs. Cement boards were attached to each side of the metal frame and foam insulation was injected into the cavity between the boards to form a well-insulated exterior skin. The largest available panel, 4.6 by 2.5m (15 by 8 feet), could be lifted by two to four people. The walls were not only strong enough to support a second storey but they could also be installed horizontally for use as the roof or as the floor of the second storey during future expansion. Their thermal insulating value was far greater than that provided by cinder block which is the material typically used in such projects. The core of the panels was also the location for the electrical conduits through which the home's wiring could be fished after assembly.

The fabrication of the panels themselves is simple and does not require specialized tools. They can even be put together on site when the scale of the project justifies investment in a prefabrication shop that also provides training and employment for local workers. The low-tech features of the panels and their simple manufacture were appealing. In particular, I liked the fact that heavy machinery would not be needed for assembly. I was concerned, however, that the prefabrication aspect might be construed by the Mexican buyers as temporary or flimsy in quality. The culture of building in Latin America implies a long process but one with materials (such as cinder blocks or stone) that project an image of solidity and permanence. The introduction of a new construction method – even one that included cement board – would raise doubts in a some minds. The design of La Casa a la Carta was in line with Mexican tradition, but the new and more efficient method of construction would be a trade-off involving a degree of trust on the part of Mexicans who wanted to be homeowners. Housing authorities would also have to accept these methods if they wanted to do something about the huge housing deficit.

We had only three days in which to construct the demonstration house (figure 9). Even though we were only putting up a demo unit, I was still concerned about the short deadline. Viewed positively, it would be a test of our ability to demonstrate construction efficiency. The panels that had been constructed in Montreal were shipped to Guadalajara by truck along with the rest of the housing components. The first thing to be set up was a temporary wood platform on the floor of the local exhibition hall. Assembly time for the wall and roof panels was only ten hours. The windows were then placed in their prepared openings, and interior partitions were built on site with steel studs and cement boards. Following assembly, the surfaces of the interior and exterior partitions were covered with stucco. The workers then turned their attention to the finishes of the unit and the electrician fished wires through the conduits. Floor covering came next, with linoleum in the kitchen, living, and dining areas, wood in the bedrooms and tile in the bathroom where a shower stall, toilet, vanity, and sink were installed. The prefabricated kitchen cabinets were put in at the same time.

The interior decor of the unit was the outcome of a scenario that was created for an imaginary housing project (figure 10). We envisioned a young family purchasing a first-time home in a low-cost, perhaps social housing development. The family, made up of

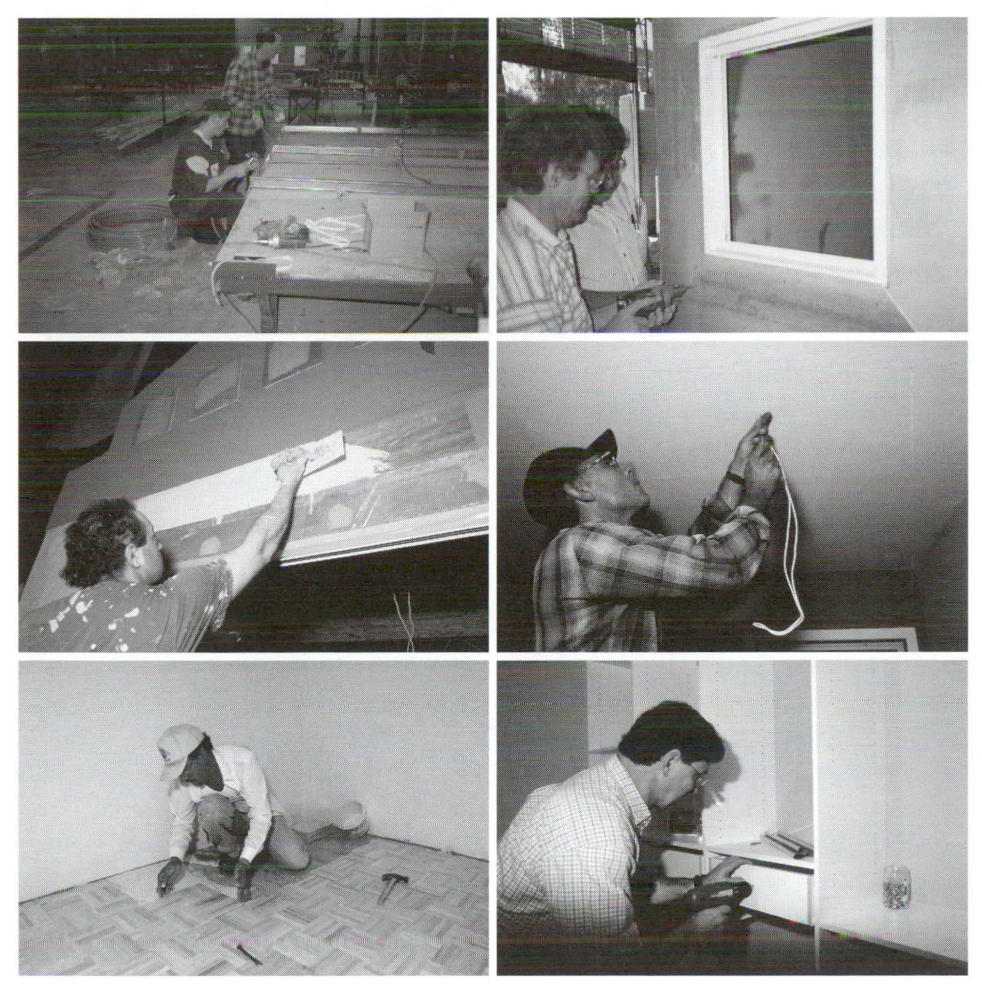

Figure 9 Construction sequence of the demonstration prototype in Guadalajara, Mexico. Clockwise from top left: Construction of the wall panels in Montreal by Archimède 2000; wall assembly on site; fishing the electrical wires through prepared conduits; installation of general storage space (eventually to be replaced by stairs); laying the parquet floor in the kitchen; plastering the exterior.

Figure 10 The demonstration unit in Guadalajara, Mexico
From top: Dining/living room area; kitchen cabinets; the small bedroom; storage space

working parents and two young children aged four and two, visited a demonstration site where they saw several model houses and menus of interior and exterior components. Based on their budget and space requirements, they made their selections. They liked the idea of the built-in flexibility which let them expand the house progressively – both horizontally and vertically – to accommodate their intentions to have more children, house a live-in relative, or supplement their income by renting out a room. The interior design reflected the colours and the furniture that one would expect to see in a Mexican home.

The house was completed on time and ready for display. I stayed nearby throughout the exhibition and spoke with many Mexican officials, builders, and potential buyers. They had trouble believing that the type and quality of finishes in the house could be made available for such a low price.

La Casa a la Carta had been launched. It did not have the same immediate effect that the Grow Home had back in Montreal, but as one local visitor explained to me, the most precious commodity needed when undertaking a new venture in Latin America is *paciencia*: patience.

9 SMALL IS GREEN

When Grow Home buyers moved into their first homes, they were content. Life as a renter was replaced by life as an owner. Soon they began to re-evaluate their new family budgets: mortgage payments, municipal taxes, house and car insurance, gas, groceries, daycare – the list grew longer. Would they be able to pay for it all on time? They were anxious to see the total on one particular bill: heating. It gets very cold in Montreal, and on some winter nights the temperature can plummet to -30°C. The Grow Homes were heated primarily by electric baseboards. Despite the fact that hydroelectric power is inexpensive in Quebec (about 6¢ per kWh, compared with 17¢ in San Francisco and 19¢ in New York City), it can still cost a lot to heat a home. The new owners recalled their bad experiences as renters when the landlord complained about how expensive it was to heat their drafty apartments. So when their own heating bills arrived, they could hardly believe how low they were: an average of only $94 per month during the heating season. What these Grow Home buyers did not know was that many of the features that made their homes affordable also made them energy efficient; that by buying and living in a Grow Home they were making a valuable contribution to the efficient and responsible use of natural resources. Whether they knew it or not, they now supported the mind-set that holds that our life on this planet must be sustainable.

The notion of sustainable development is based on the conservation of natural resources. A resource is any material that is needed or used to sustain life and livelihood: air to breathe, water to drink, land upon which to build homes and grow food, forests for timber and paper, ores for metals, oil and natural gas for energy needs. There is hardly a human activity that does not draw upon natural resources. Many believe that our present industrialized societies use too many virgin resources and degrade the environment in countless ways, and that such practices cannot continue much longer without serious consequences. Our ability to protect the environment lies in the effective management of resources that can be replenished or reused and in the efficient use of those that cannot.

Resources can be divided into three general categories (Friedman et al. 1993). *Renewable resources* are those that are replenished through relatively rapid natural cycles. Examples include water, air, and biological products such as food, fibres, and timber. In theory, a renewable resource can last forever. However, all renewable resources are limited by the capacity of natural systems to renew them. For instance, natural processes that purify water do not occur at an adequate pace to replenish clean water that is consumed and reintroduced into the environment as waste. Learning to manage and use renewable resources while respecting this limitation is the proper study and practice of conservation. The management or conservation of renewable resources involves practices that preclude over-exploitation. *Non-renewable resources than can be recycled* include all non-energy, mineral resources that are extracted from the earth's crust. Ores of copper, aluminum, mercury, and other metals and minerals are examples. As these materials are mined, they are not replaced, at least not at a sufficient pace to be relevant within the human time frame. However, people can collect these materials after they are used and recycle them. *Non-renewable resources that cannot be recycled* are mineral energy resources such as fossil fuels, including coal, oil and natural gas, and uranium which is used for nuclear energy (Nebel and Wright 1993). Humans expend in one year the quantity of fossil fuels which it took nature approximately one million years to create (Buchholz 1993). There is no way of recycling the energy in fossil fuels: once the energy has been released, it cannot be regained.

The construction and occupancy of homes consume resources and generate waste. Construction is responsible for 16 per cent of the total solid waste production, and approximately 20 per cent of this is from new homes. About 80 per cent of this waste ends up in landfills, much of which could be avoided. A study conducted in the Greater Toronto area during a period of high building activity revealed that construction of an average home produces two and a half tonnes of waste, and twenty tonnes when demolition is required (REIC 1991). Approximately a quarter of the waste is dimensional lumber, and another 15 per cent is attributed to manufactured wood products. This situation is not only wasteful in terms of embodied energy but also contributes to the problem of waste disposal. Despite Canada's vast terrain, most of its population is concentrated in a small number of urban centres, and waste disposal in landfills has become a serious crisis, with a growing problem of toxic leaching, resulting in contaminated soils and groundwater.

Energy consumption is also significant. Requirements for countries in cold climates are generally high, partly due to space heating requirements. Energy usage per capita in Canada is one of the highest in the world. Heating, cooling, and operating housing account for approximately 20 per cent of our total energy consumption. Furthermore, an additional 5 per cent to 10 per cent is used as indirect or embodied energy from construction, renovation, and demolition of housing and its infrastructure (Robinson 1991).

So how should one design with the environment in mind? Designing a home within the context of sustainable development requires that the impact of the home on natural resources be minimized. Attention must be paid to the rate at which resources are depleted and the extent to which the environment is degraded. Architects and builders need to consider a reduction in the consumption of renewable resources to allow the earth's natural cycles to make them available at a sustainable rate; the recycling of non-renewable resources to make them available for processing into new products; and the improvement of the efficiency of use of non-renewable, non-recyclable resources.

In the design of the Grow Home our intention was to lower housing cost. However, we soon recognized that the same principles that made the house inexpensive also made a significant contribution to the reduction on non-renewable, non-recyclable resources. We set out to formalize these principles and to quantify their advantages.

The first strategy that contributed to the reduction of resource consumption in the design of the Grow Home was efficient planning of the unit's interior layout. By increasing the usable floor area, the quantity of construction materials was reduced, as were the space heating requirements. The design objective was to provide a smaller house with the maximum usable floor area so as not to disrupt the occupants' living comfort. This was achieved in several ways. The open interior plan, for instance, contributed to energy efficiency and flexibility and used less material. Local heat gains and losses were easily equalized, and energy demand for mechanical ventilation was reduced since there were no obstructions to the air flow. Circulation paths and hallways, which receive marginal use, were reduced to a minimum. Planning efficiency was also increased by grouping spaces with similar functions and environmental control needs: pipe, duct, and conduit runs were minimized by planning for close bathroom, kitchen, and laundry areas with back-to-back sinks.

Future design of the Grow Home can reclaim the attic. Most types of prefabricated trusses can be wasteful where small spans are involved, particularly since they render the space unusable for living purposes. Using special trusses or rafters in the roof could increase the floor space without necessarily increasing construction costs. In a Grow Home of 4.3 by 11m (14 by 36 feet), for instance, a 2.4m (8 foot) clearance could be achieved with a roof slope of 6:12. Assuming that 40 per cent of this space is usable for occupancy, an additional 19m^2 (190 ft^2) of floor space could be added to the Grow Home. Since heat rises, the attic space would benefit from the heat of the lower floors.

One of the easiest and most direct contributions to the reduction of material use and heat loss was the simplification of the unit's configuration. A more complex building form has more corners and perimeter, which in turn requires more "skin." This results in higher construction costs and increased heat loss. Generally, the ratio of floor area to perimeter in the Grow Home was maximized. Figure 1 illustrates several possible building configurations with the same floor area. At one extreme the H-shaped plan has a ratio of floor area to perimeter of .87, requiring

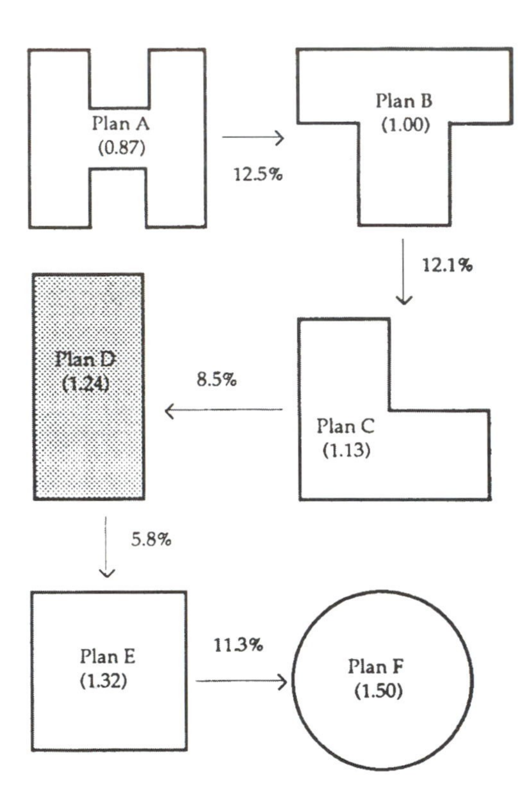

Figure 1 Effect of building configuration on perimeter and floor area (numbers in parentheses refer to ratio of floor area to perimeter)

Plan Configuration	Wall Area (m^2)	Energy Required (KWh)	Monthly Heating Cost ($)
Plan A (H)	160	2856	134
Plan B (T)	140	2501	117
Plan C (L)	123	2198	103
Plan D (rectangle)	112	2001	94
Plan E (square)	106	1894	89
Plan F (circle)	94	1679	79

Table 1 Effect of building configuration on energy consumption

160m² (1,600 ft²) of exterior wall area for a 100m² (1,000 ft²) house. The required envelope decreases progressively as the plan is simplified to a T-shape, an L, a rectangle, square, and circle. This last configuration makes the most efficient use of space, requiring almost 11 per cent less wall than the square for the same floor area.

A simple plan like that of the Grow Home costs less to build since there are fewer corners and fewer windows. Envelope costs, from the basement to the roof, were reduced since simple configurations generally require less cutting and fitting of building materials. Consequently, the amount of material wasted was reduced and the management task was simplified at the same time. Reductions in the exposed wall area were accompanied by a proportional decrease in heat loss. When a plan is simplified from a T-shape to a square, conductive heat losses from the walls alone can save $28 in heating costs annually for a small (100m² [1,000 ft²]) house in Montreal, Canada (table 1), perhaps insignificant individually, but with considerable environment impact when multiplied over thousands of units. These savings were easily doubled when the additional heat losses from the basement and infiltration were accounted for.

While the circle provides the best area-perimeter ratio, its potential savings are offset by the extra cost of building curved walls, while the interior plan of a circle could lead to layouts that are not functional. Similarly, the square provides a more efficient perimeter-to-floor-area ratio when compared with a rectangle, but this configuration may be difficult to plan efficiently on the interior. The Grow Home rectangle appears to be the most advantageous configuration in this regard. Furthermore, the rectangle requires less land, since it can be built on narrower lots and can benefit more easily from being grouped into rowhouses, which in turn leads to additional savings in construction and energy costs.

Grow Home builders have paid special attention to another valuable component in their quest for cost savings: modular design. They understood that a simple and effective way of reducing material waste was through careful dimensioning of the building to accommodate modular configuration of building materials. At the most basic level, designing within standard dimensions for structural framing members such as studs, joists, and plywood resulted in substantial savings. In a typical detached home, the use of general dimensions for stud spacing (i.e., 405mm [16 inch] module) to eliminate the need for an extra stud at the end of the wall, placing and dimensioning windows accordingly, and locating partitions to line up with the structural studs saved, altogether, a tonne of lumber per home. Designing for 1220mm (4 foot) modules and 610mm (2 foot) stud spacing reduced lumber use by 8 per cent. Providing for efficient details at corners and intersections of exterior walls and interior partitions doubled these savings. The same principle of dimensioning was implemented to accommodate interior finishes such as drywall and floor tiles. Cost savings were achieved not only through efficient use of materials but also through reduced labour since less cutting and fitting were needed.

Figure 2 illustrates four simple Grow Home plans which have approximately the same floor area and configuration but are dimensioned for different modules. Theoretically, the amount of structural wood re-

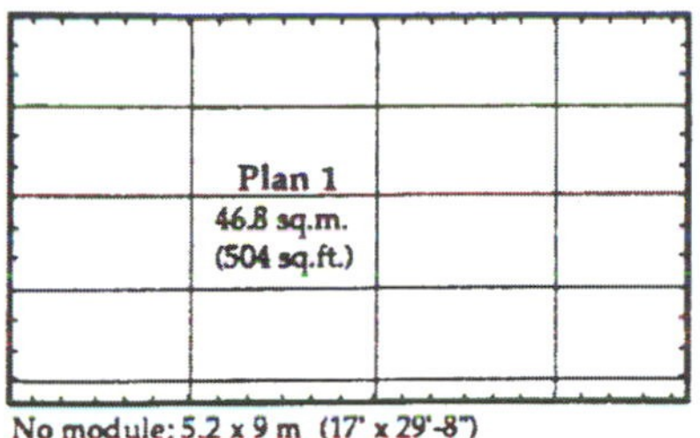

No module; 5.2 x 9 m (17' x 29'-8")

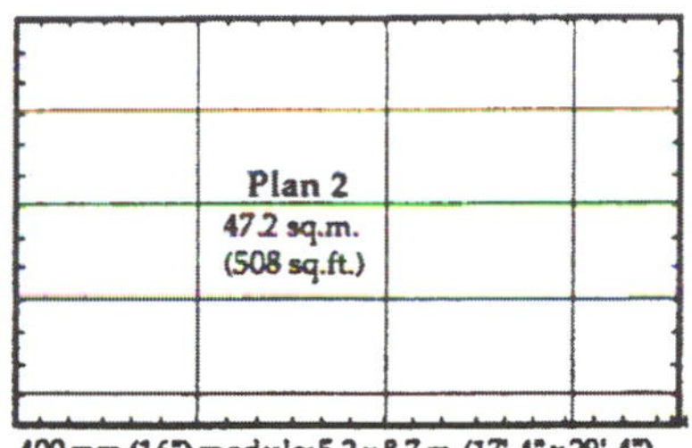

400 mm (16") module; 5.3 x 8.7 m (17'-4" x 29'-4")

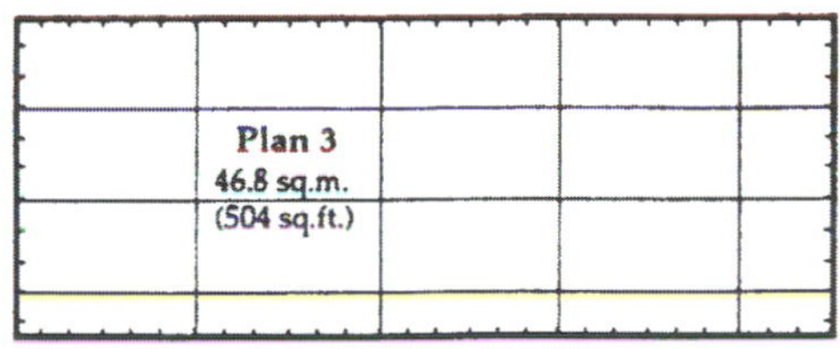

600 mm (24") module; 4.3 x 11 m (14' x 36')

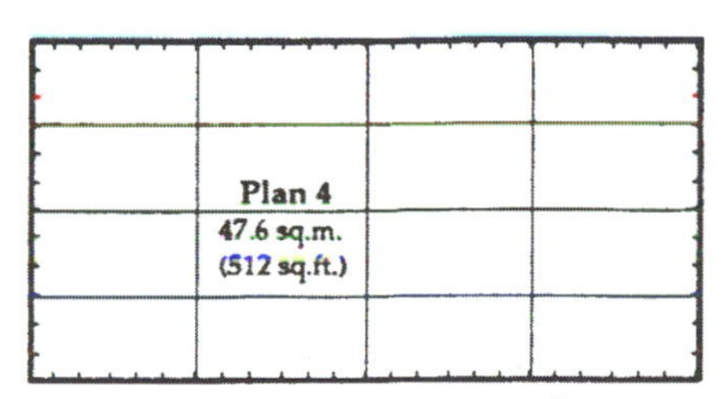

1220 mm (48") module; 4.9 x 9.6 m (16' x 32')

Figure 2 Alternative modular design/dimensions

Plan Module	Joists (m)	Studs (m)	Sheathing (m²)
Plan 1: (no module)	378.2	561.1	232.5
Plan 2: 405mm (16 in.)	384.4	520.1	233.3
Plan 3: 610mm (2 ft.)	299.9	559.2	242.4
Plan 4: 1220mm (4 ft.)	302.4	534.0	237.9

Table 2 Measured material requirements

	Joists		Sheathing	
Plan Module	Ordered (m)	Wasted (m)	Ordered (m²)	Wasted (m²)
Plan 1: (no module)	403.6	25.4	249.7	17.2
Plan 2: 405mm (16 in.)	408.4	24.0	249.7	16.2
Plan 3: 610mm (2 ft.)	302.4	2.5	249.7	7.3
Plan 4: 1220mm (4 ft.)	302.4	0.0	237.9	0.0

Table 3 Material usage

quired to build these plans should be almost the same, since there is no significant difference in either floor area or perimeter between the alternatives. Table 2 shows the quantity of floor joists, wall studs, and sheathing materials (for walls and floors) required for a two-storey house as measured from the plans. The sheathing requirement is fairly consistent from one plan to another, with a difference of about 10m² (100 ft²) between the highest and lowest estimates. This is equivalent to about three and a half standard sheets of material. Similarly, the total length of wall-framing components (including top and bottom plates) changes little between alternative floor plans except for Plan 2, which approaches a square and has less perimeter than the others. More floor joists are needed for the first two plans (plans 1 and 2), which are wider and require tighter joist spacing (405mm [16 inches]).

Although the total measured material requirements are much the same for plans using different modules, the amount of waste generated in each design can vary significantly. The amount of waste produced is generally higher in those plans that are based on smaller modules or random dimensions (table 3). Between 6 per cent and 7 per cent of the material bought for Plan 1 (random dimensions) is wasted on off-cuts that cannot be reused because they are too small. This wastage is reduced gradually as the plans are designed for larger modules. The structural frame based on a 1220mm (4 foot) module needed very little cutting, and could be built to produce little or no waste. Consequently, 5 per cent less sheathing material needed to be purchased. The amount of lumber required in the building frame could be further reduced by using framing alternatives outlined in table 4.

The stacked-floor townhouse form of the Grow Home, which had its greatest impact on land use and housing density, also had a substantial effect on the quantity of building materials used and, to a large extent, on energy efficiency. Vertical designs make the most efficient use of space, since more stacking results in the need for less construction material. The cost of a two-storey square house, for instance, is less per square foot than a one-storey with equivalent area, since it has half the foundation and roof area. Floor-to-floor heights, which are affected by such factors as the floor thickness and the presence of suspended ceilings, will also have an impact on

Framing Alternative	Lumber Savings (m)	Lumber Weight (kg)	Construction Cost Savings ($)	Embodied Energy (MJ)[1]
Lining up joists with studs	61.0	204	139	1502
Stud spacing at 610mm (2 ft.)	68.3	228	156	1682
Exterior walls	19.5	65	44	481
Interior walls	97.5	218	200	1579
Total		**713**	**539**	**5262**

1 Based on energy intensities and conversion factors from Sheltair 1991

Table 4 Savings from efficient framing practices

the amount of raw materials that go into construction, especially in the building envelope. Generally, buildings like the Grow Home with a smaller ratio of surface area to volume make the most efficient use of materials and require less energy to heat. Bungalows, which have an average surface-to-volume ratio of .38, are considered to be wasteful. Split-level-type plans, which accommodate the same floor area on one and a half storeys, have a lower ratio, usually in the area of .25.

The effect of floor stacking on energy efficiency will depend on several factors, including unit size and grouping (figure 3). In the case of the Grow Home, the energy benefit of moving from a one-storey, 100m² (1,000 ft²) bungalow to a two-storey detached model with the same floor area is marginal. Although the switch increases the total exposed wall area by more than 50 per cent, the extra heat losses are compensated for by a reduction of basement and roof areas. Total heat losses are therefore fairly balanced, resulting in negligible savings in the order of 1 per cent (table 5). By reducing the footprint of the Grow Home, however, and consequently excavation and foundation requirements, substantial construction cost savings were achieved.

For a narrow-front house like the Grow Home, the major advantage of stacking floors was that it allowed two or more units to be joined, leading to significant reductions in heat loss and improved land use efficiency. One of the most effective ways of reducing energy consumption in Grow Home developments was by joining units into semi-detached or rowhouse configurations, since heat losses would be limited to two walls (or three, for a semi-detached unit) and a small roof area. When compared with a completely detached unit, the semi-detached and rowhouse versions of the Grow Home offered reduction in heat losses of 21 per cent and 43 per cent, respectively. Annual heating costs were estimated at $752 for a semi-detached unit and $564 for the rowhouse (Friedman 1991). Grouping units was also an effective way of improving construction efficiency.

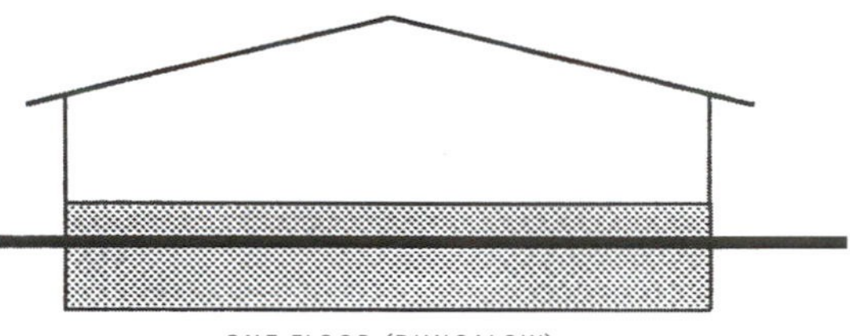

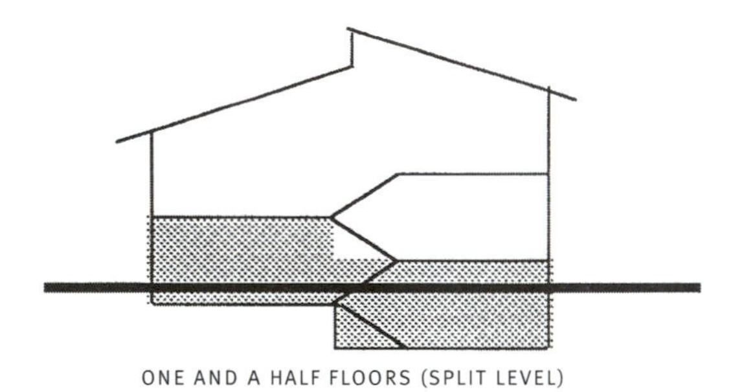

TWO FLOORS (COTTAGE)

Figure 3 Possibilities for floor stacking

Component	Heat Loss (Watts) 1 Storey Bungalow	2 Storeys Detached	 Semi-det	 Rowhouse
Roof	558	279	279	279
Walls	1005	1643	1002	367
Doors and windows	1598	1598	1598	1598
Basement	1560	1249	780	311
Infiltration	1547	1413	1250	1087
Total	**6267**	**6182**	**4912**	**3642**
Annual energy consumed (KWh)	9154	9029	7174	5320
Associated heating costs ($)	429	423	336	249

Table 5 Combined effect of floor stacking and unit grouping on heat loss

The reduction in perimeter area also had a significant impact on the delivery time, since construction of the envelope is a labour-intensive operation. The joining of units into groups of two or more provided significant savings in both construction and energy. Joining four detached units into semi-detached, for instance, reduced the exposed wall area by 36 per cent. Grouping all four units as rowhouses provided an additional 50 per cent savings (figure 4).

Heat-loss reductions of approximately 21 per cent were achieved when two dwellings were attached, and a further 26 per cent saving for the middle unit when three or more dwellings were joined as rowhouses (table 5). For Montreal, the difference between a detached unit and a rowhouse is an annual saving of $118. In addition to substantial reductions in energy consumption, the joining of units enabled efficient use of land and infrastructure.

The prime intention of Grow Home builders was to attract first-time buyers with modest means. They lowered the unit cost by building it small. In so doing, they also reduced demands on many depleted resources such as lumber. This type of contribution is not unique. In recent years, the public at large has become more aware of environmental concerns. Terms like global warming, phenomena such as sudden mid-season temperature changes, and natural disasters around the world have made people more aware of the subject. Scientists now acknowledge that events on earth are contributing to these phenomena. The depletion of the ozone layer, for instance, is now attributed to carbon monoxide emissions. Common citizens are now willing to take proactive steps to remedy the problem. Recycling programs, for example, have been implemented in many cities across the continent. Some of these concerns have also

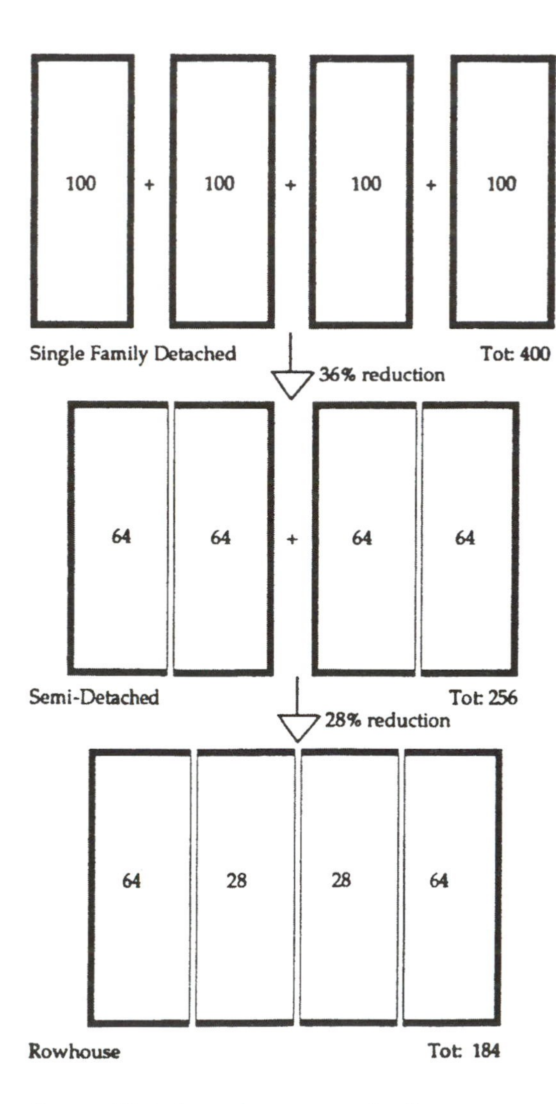

Figure 4 Effect of grouping on exposed wall area

found their way into the educational curricula of schools at all levels. Governments and trade associations promote change in building practices, and they advocate energy conservation and efficiency. Home buyers are encouraged to purchase better-built and better-insulated homes. The underlying trend in home-building and general consumption, however, runs counter to all of these good intentions, and that is the trend of building big. Grow Home buyers bought small for economic reasons but in the process also made a contribution to saving the earth's resources.

It was a chilly spring morning in Meerlo where I had arrived the previous night to attend a conference. I was up early because of jet lag and left my hotel to walk around this Dutch neighbourhood on the outskirts of Eindhoven (figure 1). The streets were mostly empty. From time to time, someone stepped out of one of the townhouses that lined the streets, offered me a greeting, then got into a parked car and drove away. The houses were clad with brick and had red clay tiles on their roofs. The facades were simple, some painted white. Door frames were lined up with windows on the upper floors, and most of the windows had wide black flower boxes fastened to the sills. Many residents had also hung ornaments on the inside of their windows. The streets slowly began to fill with cars and bicyclists, schoolchildren waving to their parents as they set off to the neighbourhood school, well-dressed men and women lining up at a bus stop near an intersection, and older people pausing to speak with one another outside a corner bakery fragrant with the smell of freshly baked bread.

I looked again at the homes. It appeared that Meerlo had been built after the Second World War. In between the newer homes were recognizably older ones. Maybe it had been an old village that expanded over the years. The homes were not identical, yet there was a definite sense of harmony in the ensemble. They were close to the narrow public sidewalk, perhaps set back two metres (six and a half feet), leaving a narrow strip for meticulously tended private gardens with many plants and bushes. I wondered how such a small strip of land could appear so striking. A special path of red unipavers alongside the narrow road was clearly marked as the lane for cyclists. At the intersection, the street elevation was raised to slow traffic and to let people cross safely. Special bays shaded by trees were marked for public parking. Nowhere could I spot any indoor garages, just the occasional carport between groups of

Figure 1 Meerlo, a neighbourhood on the outskirts of Eindhoven in the Netherlands

homes. Small public parks were provided every few blocks, green patches with play structures in them. The lower floors of some of the townhouses had stores at street level, with the residential quarters on the upper floors. In one grouping that was slightly recessed, I noticed a grocer, florist, optician, and post office, with several cars parked in the bay out front.

Walking the streets of Meerlo was a pleasant experience. In marked contrast to the wide streets and lower density of back home, where each home strained to be different from its neighbour, there were no three-car garages or oversized lamp posts. Meerlo was not an affluent community, yet it reflected a sense of elegance. I wanted to understand what made walking there so comfortable. Was it the architecture? The wonderful gardens and flower pots? The cheerful greetings and smiles of the passersby? It dawned on me that the combined effect of everything I saw on the street – as well as the scale in which it all existed – was what gave the place its charm. Scale in particular was the elusive ingredient that had disappeared from the North American urban landscape. Human scale was evident in the proportion of many features I saw that morning in Meerlo, and it created a sense of being part of the big picture.

I often recalled the images of Meerlo and other European towns when I was asked to design communities using the Grow Home. My clients were merchant builders operating against a number of disadvantages. First, they were not always welcome. Many citizens with not-in-my-backyard sentiments argued that affordable housing would lower the value of their properties. There also existed lower-density zoning bylaws that favoured detached homes with large setbacks and wide boulevards. Changing these bylaws is an uphill battle destined to failure. There were no municipal allocations of land for small parks or zonings for a mix of homes and stores. With not enough room on the street in a high-density community, parking would have to be in lots and, in many cases, indoor garages. On the builders' side, budgets were tight. Small amounts had been devoted to landscaping and street furniture and to facade and roof articulation. I knew that a proposal would always have to be made to a municipality and that a grueling process of negotiation would follow. I would also have to pressure the builders and hope that they would loosen their purse strings.

Such was the case in Aylmer, Quebec. The developers Denis Lefebvre and Jacques Charrons, owners of JAG-DO Construction, had heard about the Grow Home success in Montreal and wanted housing based on similar principles for their own site. The 3.24 hectare (8.1 acre) site, adjacent to the Ottawa River in an area called Deschênes, was thought to be a good location for federal employees, mostly first-time buyers, who worked across the river in the capital. The initial public response to the introduction of affordable housing in Aylmer was resistance. We persisted and developed ten schemes between August 1991 and March 1992, and also made numerous presentations to councils and committees. We resisted pressure to lower density which would have increased housing cost, and we kept the Grow Home principles intact. During meetings with representatives of Aylmer's planning department and with the city's citizen committee, we argued that development standards must be changed to allow affordable housing. The required unit ground coverage was reduced from 50 to 47.3m² (500 to 473 ft²), the front setback from 7.5 to 5m (25 to 16 feet), and the width of the pedestrian path from 3 to 1.5m (10 to 5 feet), thereby saving thousands of dollars per unit. Today Le Quartier du Parc Madaire is a thriving community. The 115 units were sold in two months for $68,000 each.

Despite such difficulties, I continued to embrace the challenge and over the years, with assistance, I designed upwards of a dozen communities. I realized that the main challenge for affordable narrow-front rowhouse developments – with "squeezed space" by North American standards – was to make them pleasant and livable. In addition, the social stigma attached to this type of housing, especially in established communities where the single-family detached home predominates, may be overcome if the denser communities – often associated with barren and sterile surroundings – are designed with forethought and care. I identified three factors necessary to achieve pleasant environments: parking and vehicular circulation, private and public open spaces, and unit and community identity. I concluded that these are the ingredients which, when properly manipulated in the design of a community, can instill a sense of scale along with the proportions of the built environment itself (figure 2). These factors are described below.

The extensive ownership and use of the passenger car along with the vast network of public roads has promoted the phenomenon of leap-frogging, a pattern whereby builders, because of lower costs, develop land increasingly further from town centres. New affordable communities are almost always located on or beyond the urban fringe: the potential benefits of such developments are lower home prices due to reduced land costs and the relocation of the labour force closer to employment centres which have been moved out of the city cores. The disadvantages, however, include urban sprawl, higher transportation costs resulting from increased commuting distances, and a greater dependence on the car which aggravates the associated problems of automobile emissions, traffic congestion, and parking. Whether in an urban or suburban setting, the car is an inescapable reality in affordable communities. Parking in a project of forty-five to sixty units per hectare (eighteen to twenty-four units per acre) can account for nearly 50 per cent of the total site area. The higher the density of a development, the greater will be the impact of parking and vehicular circula-

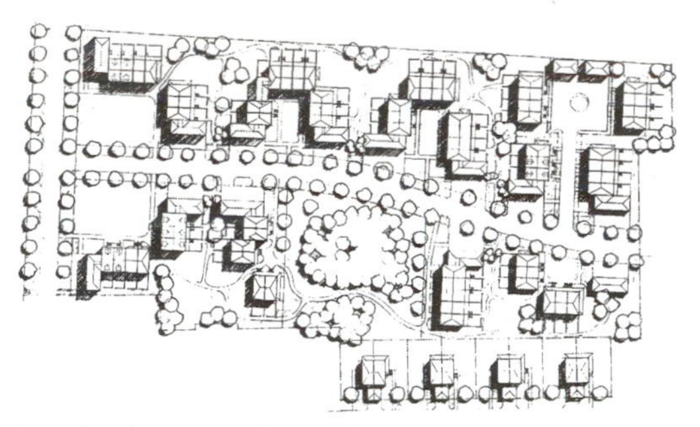

Quartier du Parc Madaire, Aylmer, Quebec – Alternative 1
Site area: 3.24 hectares
Number of units: 75 Gross density: 23 units/hectare
Shared covered parking: 67 Private covered parking: 8

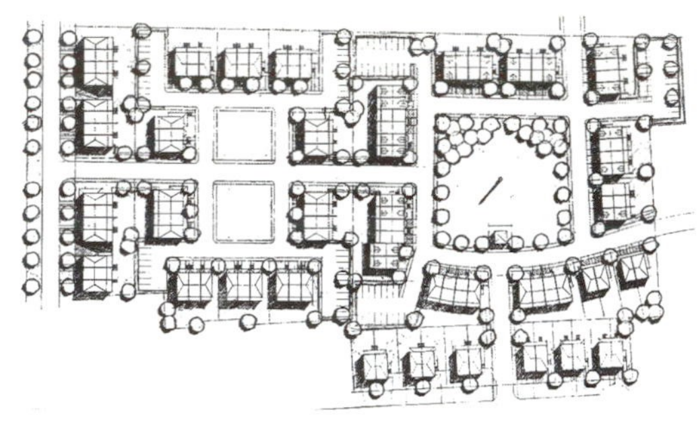

Quartier du Parc Madaire, Aylmer, Quebec – Alternative 2
Site area: 3.24 hectares
Number of units: 119 Gross density: 37 units/hectare
Shared covered parking: 79 Private covered parking: 60

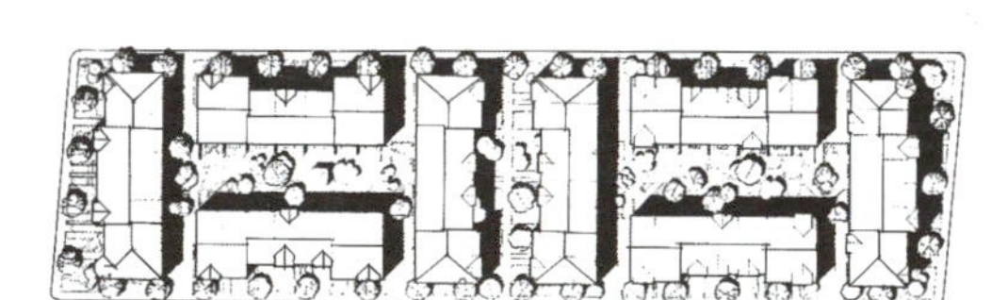

L'Îlot de Marseille, East End Montreal, Quebec
Site area: 0.69 hectares
Number of units: 60 Gross density: 87 units/hectare
Shared covered parking: 48 Private covered parking: 12

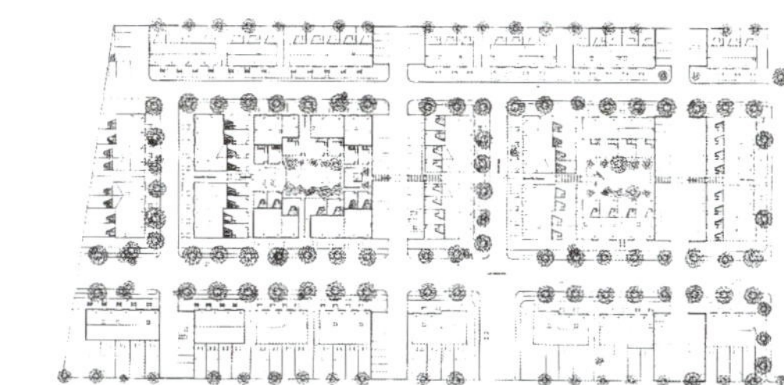

Quartier du Parc Vinet, East End Montreal, Quebec
Site area: 2.7 hectares
Number of units: 122 Gross density: 60 units/hectare
Shared covered parking: 137 Private covered parking: 58

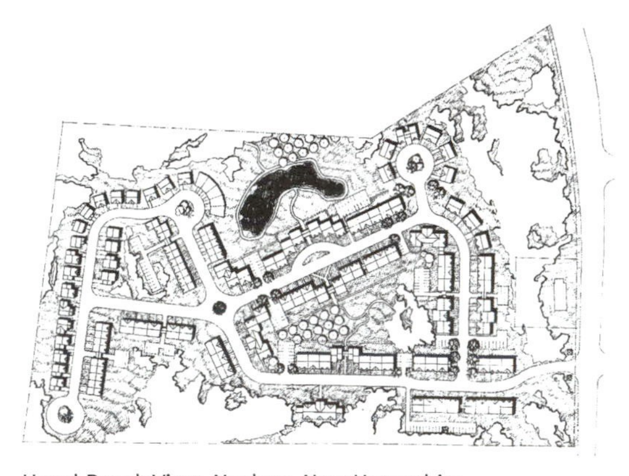

Hazel Brook View, Nashua, New Hampshire
Site area: 10.95 hectares
Number of units: 220 Gross density: 20 units/hectare
Shared covered parking: 250 Private covered parking: 36

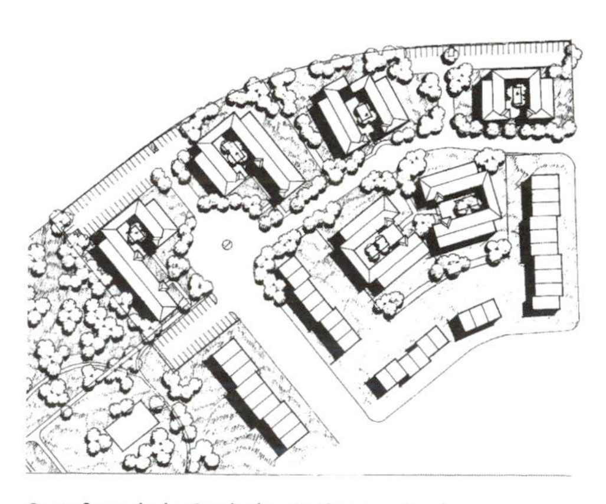

Carrefour de la Capitale, Gatineau, Quebec
Site area: 1.9 hectares
Number of units: 55 Gross density: 29 units/hectare
Shared covered parking: 99

Figure 2 Narrow-front affordable communities

Figure 3 Parking under the rear balconies in the Cité Jardin Fonteneau development in Montreal (Cardinal and Hardy, architects)

tion. It is therefore of utmost importance in high-density developments to treat parking in an efficient and unobtrusive manner. Providing indoor garages consumes valuable unit space and creates long expanses of asphalt in front of the row.

The visual impact of the car (i.e., very wide roads, large parking lots, long series of repetitive garage doors) can be reduced when parking is integrated into the landscaping to diminish its apparent presence (figure 3). Several smaller screened parking areas result in less of a visual presence than one large parking lot, as can be seen in the site plan for a development in Nashua, New Hampshire, where parking was relegated to the rear of the units in a number of small lots surrounded by landscaping (figure 4). Depressing the parking areas or berming their perimeter, combined with appropriate landscaping, are effective methods of concealing them. When sites for affordable communities are marginally located, parking areas can be

L'Îlot de Marseille: In an infill city block, interior shared and private parking were provided in order to reduce the number of garage doors

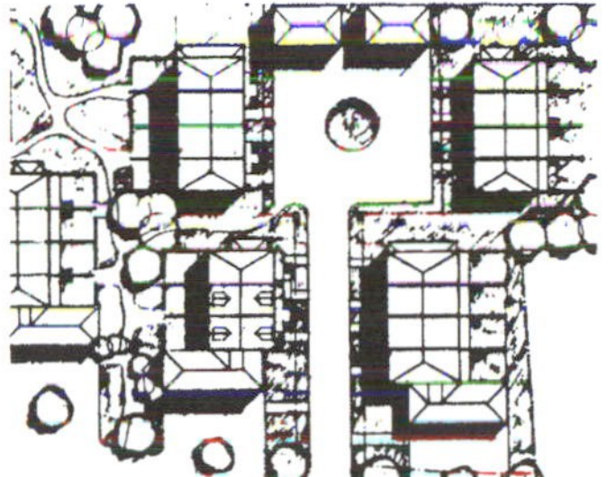

Parc Madaire: Covered parking structures were placed off the main road and in the rear

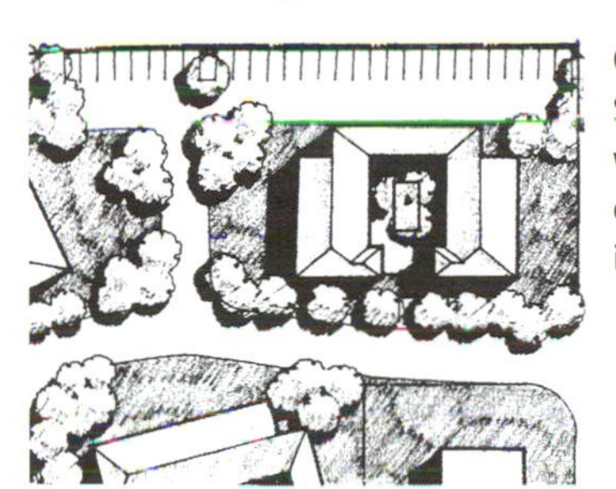

Gatineau, Quebec: Shared surface parking was placed at the edge of the property, bordering a busy highway

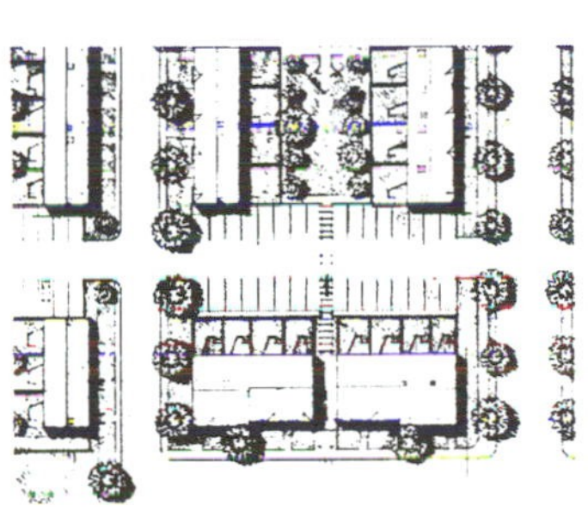

Parc Vinet: Shared surface parking was placed within a short distance of the units

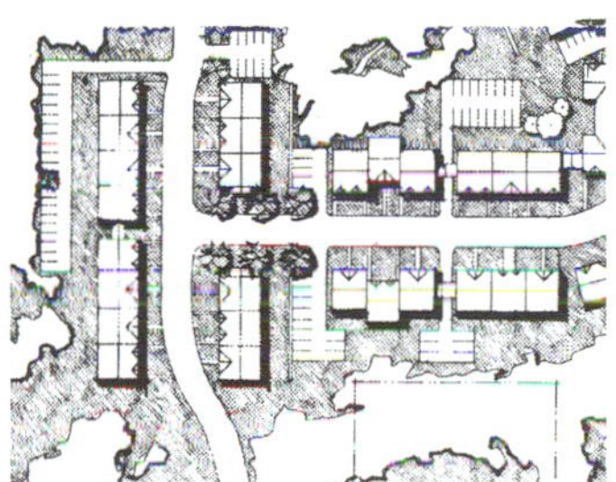

Nashua, N.H.: Parking was pushed to the rear of the units in order to reduce visibility from the main streets

Figure 4 Parking alternatives in narrow-front affordable communities

used to separate the housing from unattractive adjacent elements, as in the case of the Gatineau development where shared surface parking was located at the edge of the property that bordered a busy traffic artery (figure 4). The strategy of paving with textured blocks instead of asphalt not only increases the visual effect but absorbs storm water, thereby reducing the infrastructure required for storm runoff. Van der Ryn and Calthorpe (1986) advocate the centralization of parking in an underground facility to reduce visible paved surfaces with their inherent construction costs and storm drainage and also to enhance the pedestrian quality of the neighbourhood by discouraging the use of the car within the development. Cooper Marcus and Sarkissian (1986) suggest that determining the level of parking per household is an essential element in the initial design of the community; this level is estimated according to such factors as current rates of car ownership, the life cycle stage of the potential inhabitants, their socioeconomic status, the quality of public transport, and the general availability of the site area for parking.

Vehicular circulation in high-density communities often creates conflicts with pedestrian circulation and play areas for children. Narrowing street width and establishing a clear hierarchy of priorities not only reduces costs but can improve safety by slowing down automobile speed. Designing parking areas on the periphery of the developments leaves the core of the site vehicle-free (Cooper Marcus and Sarkissian 1986). The use of speed bumps, cobblestone segments, and highly textured driving surfaces such as stamped concrete and the emphasis of entryways by the placement of gateways are useful strategies for controlling vehicular speed. In the Quartier du Parc Vinet project (figure 4), the City of Montreal allowed narrower street widths which contributed to lower unit prices as well as to the level of safety. The shared surface parking at Parc Vinet was concentrated in a number of small areas, screened with fences and landscaping, and was located within short walking distance of the housing units.

The various forms of tenure suited to the narrow-front rowhouse community include freehold, co-ownership, and condominium. In freehold tenure where individual residents each own their unit and lot, and in co-ownership tenure where a group of residents enters into an agreement to share ownership of their units and lots, the public space accessed by all residents is owned by the city. In condominium tenure the residents own the structure of their respective units while the lots and common open spaces are owned in unison. Where the access routes of a rowhouse development are narrower than the standard required by municipal zoning, they are designated as private roads and owned conjointly by the residents, an arrangement suited to condominium tenure. Strong community identity and an equitable shared use of common open space are frequent results of condominium tenure in a rowhouse development.

When personal space is diminished in a rowhouse community, communal space takes on an added significance to the visual and functional stimulation it already provides. Some of the essential elements to achieving successful public spaces which accommodate a variety of activities are established levels of privacy, a clear demarcation of edges, benches, landscaping, and hardscaping. The clear distinction between private and communal open areas is of utmost importance. Kevin Lynch (1990) maintains: "Careful manipulation of the edge and the access system is the key to design." Cooper Marcus and Sarkissian (1986) stress the delimitation of the private from the public in high-density developments, emphasizing that differentiation is "especially necessary where private open spaces abut onto a communal landscaped area."

The front yard is significant in the rowhouse community since it provides both a transition zone between the private and public realms of the house and a link with the social fabric of the neighbourhood. A clear definition of front yard ownership combined with its status as a location where residents can interact with their neighbours embodies this transition zone and link. Even when the building is pushed forward to accommodate a larger backyard space, the identity of the front yard can be maintained with defining landscaping and/or fencing. The demarcation provided by the front entrance of the home can be achieved with a step, porch, or other carefully selected detailing. Where private open space in the front of the property is highly limited, balconies affixed to staggered (i.e., terraced) units provide valuable outdoor areas. In the backyard, visual privacy is achieved with hedges, fences, screens, and trellises which offer an important sense of enclosure for personal activities and domestic chores. Where patios or decks are available, sliding glass doors provide a direct link to and extension of the kitchen or living rooms. The backyards themselves, although small, are enhanced by the variety obtained through creative landscaping and covered patio space, integrated with an available facility for the storage of outdoor equipment. Microclimate is another consideration in the design of backyards: shelter from the wind and snow and a careful balance of sun and shade provide orientations that extend seasonal use. The Parc Vinet project was designed so that each unit would have its own fenced private backyard in addition to a communal landscaped area. The L'Îlot de Marseille community offered a similar benefit, with fences marking the border between private and semi-public domains (figure 5).

Any reduction in private open space can be compensated by large public open areas. Shared spaces such as neighbourhood greens, squares, and community gardens provide social gathering points and contribute to community identity (Van der Ryn and Calthorpe 1986). The proximity of public open space to the rowhouse units is important: "Access is a matter of psychological, as well as physical, connection. An open space must seem to be close and easily reached, which is very much a matter of design" (Lynch 1990). Larger public areas can serve to alleviate the apparent pressure caused by the concentration of taller structures, as in the second alternative of the Parc Madaire community (figure 5). When such a strategy is not required, a series of interconnected smaller spaces of varying appearance and shape is often preferable to one large open area (Cooper Marcus and Sarkissian 1986). The units of Cité Jardin Fonteneau are arranged in clusters, around courts (figure 6). Pedestrian paths were designed between the cluster and a communal vegetable garden was included in the project. The housing development

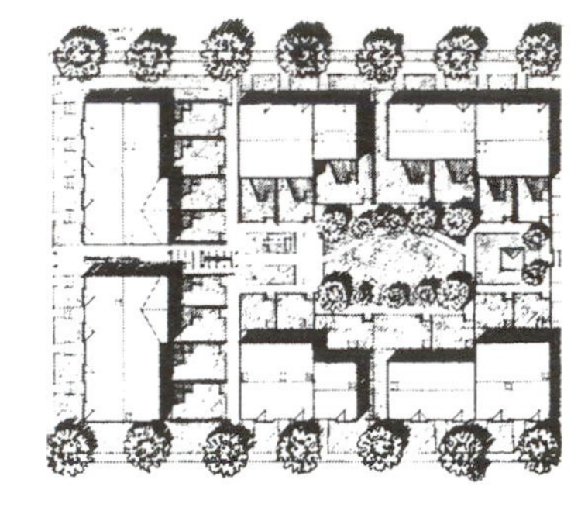

Parc Vinet: In addition to common semi-public space, a fenced private space was provided

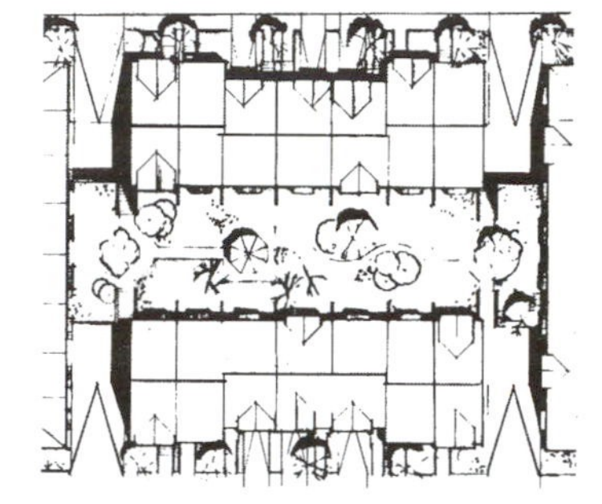

L'Îlot de Marseille: Fences indicating private domain were provided within a common semi-public space

Gatineau, Quebec: Townhouses were clustered to form a small common front yard but large private backyards

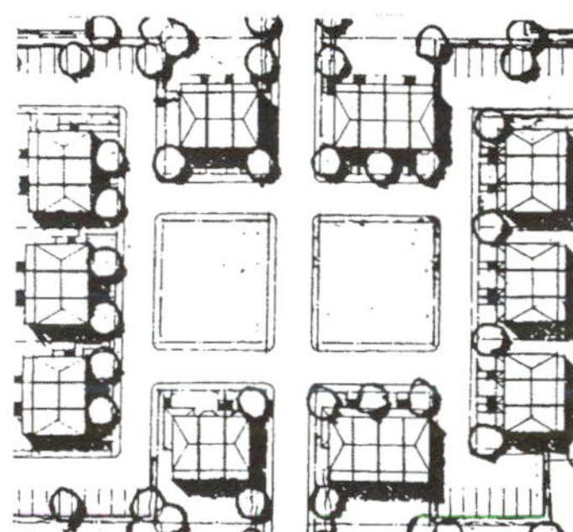

Parc Madaire – Alternative 2: The placement of narrow-front units in front of a large open space provided a wide sense of openness

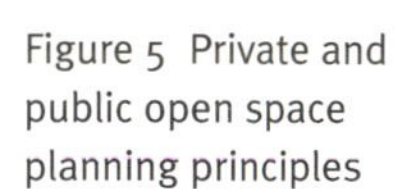

Figure 5 Private and public open space planning principles

was designed by Cardinal and Hardy Architects and built as part of a design-build competition initiated by the City of Montreal using Grow Home principles. Where rowhouses with larger private backyards are clustered around a common front area, as in the Gatineau project (figure 5), the provision of extended personal space in the rear compensates for a smaller public area in front.

In order to lower costs in high-density communities, builders rely on the ease of repetition to which the narrow-front rowhouse type lends itself. The ensuing risk of blandness resulting from of such repetition can be avoided if the designer conceives of and provides identities for both the unit and the community in the initial design. When adequate provisions are made in the primary phases of design, the desired objectives can also be achieved economically. Moreover, if the built community is appealing, increased sales and buyer satisfaction will prove profitable for the builder (figure 7).

The high degree of repetition required for economy at the level of the individual unit can be alleviated by ensuring that a set number of variable elements can be combined in novel ways to create the impression of diversity and personalization. In the Parc Madaire project in Aylmer, unit identity was enhanced by varying the dormers, porches, and facade materials (figure 8). In the L'Îlot de Marseille project, the facades were carefully articulated to achieve variety and to admit a great deal of natural light into the units; several plan options were proposed within the building shell to provide diversity for different household types and to accommodate the needs and tastes of future buyers. Even though the designer may be

Figure 6 Common court in Cité Jardin Fonteneau development

restricted for reasons of economy to a limited number of exterior components, the rearrangement and combination of these features in a creative manner can lead to interesting variations. At the unit level, a traditional approach is well suited to the design of modest, comfortable rowhouses. Tradition as a central concept in community design is essential to the neotraditionalists Duany and Plater-Zyberk: their philosophy involves the reuse, revitalization, and improvement of existing forms (Krieger 1991). In the case of the Grow Home, a classical style means the designer is not required to use odd materials or unusual shapes (which are generally expensive), and it allows for an aesthetic element within the context of straightforward construction. The positioning of openings and the choice of pleasing proportions and decorative elements "can lend even a simple dwelling a satisfying air" (Rybczynski et al. 1990).

Community identity is established through evolution and a slow process of accretion, but the conditions for such a process to occur can be provided in the initial design (figure 9). Cooper Marcus and

Figure 7 Variety of facades in the Bois Franc development in St Laurent, Quebec: clients were offered a choice of facade configuration and brick colour.

Sarkissian (1986) maintain that the general exterior impression of the community "significantly affects how residents feel about their homes, sometimes even how they feel about their own worthiness as human beings." Their approach allocates a considerable proportion of the design budget to landscaping and site amenities, even at the expense of limiting the budget on interior finishes, in order to provide "a quality milieu." The locating of trees and variation in communal outdoor areas are vital considerations, while the sequencing of views creates interest at the scale of the overall site by punctuating the design to avoid dullness. In the Parc Madaire project, where the goal was to create a high-density affordable community with the amenities found in suburban low-density areas, neighbourhood identity was emphasized by the placement of two entrances at the ends of the central boulevard; this axial boulevard underscored a connection with the existing community. In the same project the houses were designed in cluster form, each grouping with different colours and facade designs. Housing types of various footprints were arranged in the Parc Vinet project to create a well-articulated and interesting streetscape as well as to define outdoor living areas. The amount of attention paid to the overall community aspect of the rowhouse development cannot be overemphasized. As Peter Calthorpe (1993) writes, "A strong sense of community, participation, identity, and conviviality is important to support a sense of safety and comfort within a neighbourhood."

It takes time for a community to become the kind of place to which one wants to return again and again. The passage of years makes the homes more person-

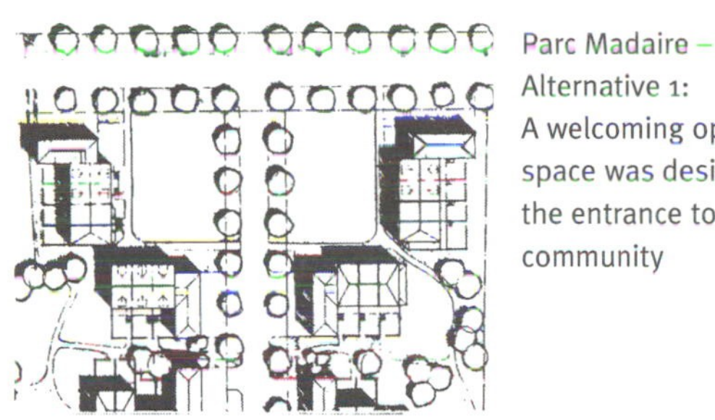

Parc Madaire – Alternative 1: A welcoming open space was designed at the entrance to the community

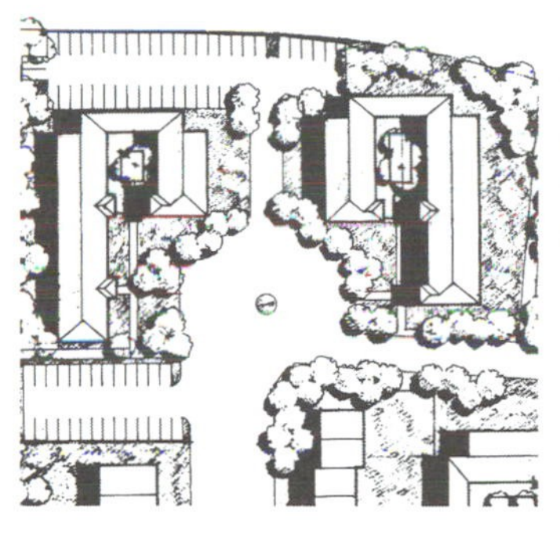

Gatineau, Quebec: The interior court of each cluster was provided with its own design features

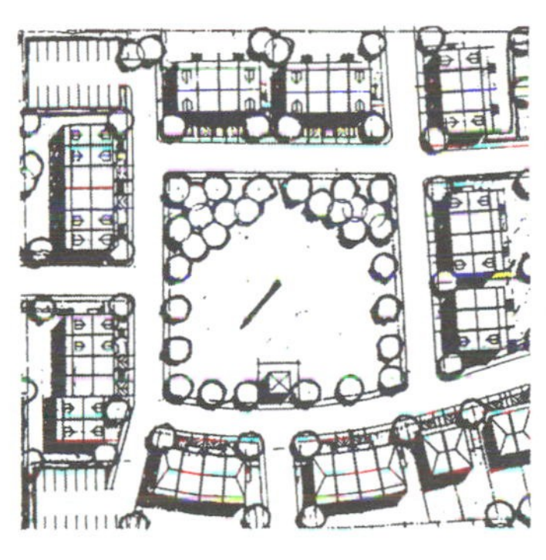

Parc Madaire – Alternative 2: An identity-forming feature such as a gazebo or flagpole was placed in the public park

Parc Madaire: Facade features were adopted for each unit to enhance unit identity

Figure 8 Identity-making features at the community and unit scales

Figure 9 Typical facades of Montreal duplexes

alized and graceful. The trees and the bushes grow taller, and a sense of scale takes hold. When basic planning principles are put in place from the beginning, the opportunity for a community to make a person feel comfortable stands a reasonable chance of success. These principles provided me with the tools to recreate the comfort that I felt when I walked through the streets of Meerlo. I hope the residents of the new communities and their visitors feel the same way.

11 WHAT'S NEXT?

I was once told by a former professor that the multi-faceted nature of an idea or concept was a mark of its richness. The Grow Home, I believe, was such an idea. It provided the opportunity for further exploration that generated new ideas. The seed components of the design were not new; they had been investigated and built before. The narrow-front townhouse has been prominent in England and elsewhere from medieval times, and it has evolved since. The design of small homes for the efficient use of space and for ongoing modifications or expansions by the residents was also not unknown in both developed and developing countries. Many of these aspects were featured in homes of the immediate postwar period. What we attempted to do was place these ideas in a contemporary context, updating them to the needs of today's society.

The Grow Home can be seen as the overlapping of three layers. The first is societal. Recent demographic transformations have given rise to the non-traditional family and new lifestyles that compel designers to introduce an increased measure of flexibility to the home. A house without load-bearing partitions enables buyers to subdivide their basements according to their diverse needs. One motive for buying the house and finishing the basement was directly related to the second layer that influenced our enterprise: the economy. The volatile employment market and the disappearance of job security as well as a sharp increase in housing prices have made the purchase of a house a challenge for many. Several would-be first-time home buyers were locked out of the home-ownership dream. The need to re-examine proven cost-reduction strategies was evident and necessary for economic reasons. The third layer was the environment. An important feature of our work was the need to recognize and be aware at all times that every design decision bears environmental consequences both during construction and after occupancy. Small size meant less spending on materials and costly heating expenses.

Exploring these three layers provided a comprehensive understanding of the full effect of the Grow Home. But it was the transformation of the Grow Home from a campus concept to actual homes on building sites that provided the greatest surprise. Many housing concepts have generated new thinking because of the sheer strength of the ideas, but not many gain the confidence of merchant builders and the buying public. The Grow Home – at least in Montreal – won this crucial vote of confidence. Through conversations with builders and visits with occupants, the built units enabled us to investigate the aftermath of the design. The Grow Home ideas generated research into flexibility and the accommodation of buyers' needs in the design of affordable housing. It also supported research into prefabrication and the personalization and choice of interior layouts and facades. The work on community design led to an exploration of planning patterns and human scale in urban environments. Environmental concerns resulted in the pursuit of a green Grow Home and the efficient use of natural resources.

In the years following the campus demonstration and the passing of the Grow Home into the marketplace, I was often asked, "What next?" The richness of the Grow Home concept continued to inspire me and suggest subsequent ideas. The genesis of the Next Home was a result of reflection on information, observation, and conclusions drawn from earlier work. The notions of flexibility, adaptability, and choice were at the heart of the new design concept. Once again a comprehensive recognition of societal changes, economic constraints, and environmental necessity served as a foundation.

The seed idea of the Next Home resulted from of the realization that the heterogeneity of the current demographic make-up, as reflected in the Grow Home, demands an approach that offers a choice between a larger or smaller unit. Unlike the Grow Home, where three levels measuring 50m^2 (500 ft^2) each were sold to a single household, the Next Home offers the possibility of purchasing a single level only. The area of each level has increased, however, to 80m^2 (800 ft^2): the structure's dimensions are 6.1 by 12.2m (20 by 40 feet), and each vertical slice can therefore be arranged differently. The three levels of a Next Home can become a triplex and house three separate households. The adjacent slice could be a duplex where elderly parents live on the lower level beneath a married son or daughter occupying the upper two floors. Next door, all three levels of the unit could be the home of a single household. The key to the flexibility of arranging the floors resides in the kit-of-parts stair idea (explored in the study on prefabrication in chapter 7). We figured out that each level would sell for $60,000 in Montreal (including land), furthering the accessibility of the unit to single-income earners. We envisioned that the flexibility would enable extended families to live together.

Another intriguing issue that arose from earlier explorations was the offering of choice in the interior layout. The notion of diverse groups of buyers with individualized space requirements demanded an approach that would allow separate arrangements according to the space needs and budgets of each household. The narrow-front design provided an advantage, enabling the liberation of space from load-bearing partitions. The invention of space joists with open webs provided an opportunity not only to place partitions wherever one wanted but to avoid the necessity of stacking bathrooms and kitchens. The development of flexible plastic conduits for hot and cold water permitted buyer choice of location not only for the living room and bedrooms but also for the utilities. The kitchen and bathrooms could therefore be placed wherever the householder wanted, regardless of the layout on the floors above and below.

To facilitate choice, we developed a detailed catalogue of interior components. The catalogue would be used first by the builder to determine the choices offered and then by the buyer. Along with the catalogue we developed a software application that would relate the choice of components to price and layout options. One could choose, for instance, to have a small L-shaped kitchen rather than a large U-shape and to locate it at the front of the unit to take advantage of sun exposure and energy savings.

To allow greater personalization, we proposed a system to include buyer participation in the choice of facade features. This was also an extension of our prefabrication study. Improving the environmental performance of the house was another lesson learned from our environmental research. The careful choice of materials and increased attention to detail were meant to contribute to this green aspect of the design.

The Next Home demonstration unit was constructed in August 1996 on the same spot as the Grow Home on the university campus. We designed a triplex to appear as though it was at the end of a row of similar Next Home structures, and to illustrate the design concept, created three imaginary households. The owner of the ground-floor unit was a widower, age fifty-seven, who had worked as a civil servant for many years and been offered an early retirement package due to budget cuts. He opted for a home office in his front room, and the furniture in the rest of the unit served as space dividers. The owners of the second-floor unit were a young couple without children. In addition to the basic area of their unit, they purchased a room attached to the back and left unfinished for completion later. The owner of the unit made up of the third floor and mezzanine at the top of the house was a single mother, age forty-one, with two school-age children. The mezzanine served as her domain and contained a study, living room, bedroom, and bathroom.

The Next Home has found its way into the market and caught the notice of the public and of builders. The choice and flexibility provided by the design attract the attention of those who believe that housing should be marketed in a new way and be affordable to all.

REFERENCES

Ahrentzen, S. 1991. "Overview of Housing for Single-Parent Households." In *New Households: New Housing*, edited by K.A. Franck and S. Ahrentzen (New York: Van Nostrand Reinhold).

Architects' Journal. 1975. Vol. 161, no. 4 (January): 182–5.

Architectural Forum. 1951. "What Next in Prefabrication?" Vol. 94, no. 5 (May): 136–7.

– 1950a. "Levitt's 1950 House." Vol. 92, no. 4 (April): 136–7.

– 1950b. "Tri-Level House." Vol. 92, no. 4 (April): 142–3.

– 1949a. "The Cape Cod Cottage: Part 2." Vol. 90, no. 3 (March): 100–6.

– 1949b. "Builder's $6,350 House." Vol. 91, no. 3, (August): 86–7.

– 1948. "House in Florida." Vol. 89, no. 1 (July): 97–103.

– 1946a. "Expansible Bungalow." Vol. 85, no. 1 (July): 116–21.

– 1946b. "Houses Off the Line." Vol. 85, no. 4 (October): 10.

– 1945. "Mr. and Mrs. McCall Know What They Want." Vol. 82, no. 4 (April): 102.

Baillie, S. 1990. "Dwelling Features As Intervening Variables in Housing Satisfaction and Propensity to Move." *Housing and Society* 17, no. 3: 1–15.

Baristow, D. 1985. *Opportunities for Manufactured Housing in Canada*. Ottawa: Canada Mortgage and Housing Corporation.

Barrett, F. 1976. "The Search Process in Residential Relocation." *Environment and Behavior* 8, no. 2: 169–98.

Beck, R.J., and P. Teasdale. 1977. *User Generated Program for Low Rise Multiple Dwelling Housing: Summary of a Research Project*. Montreal: Université de Montréal.

Becker, F.D. 1977. *Housing Messages*. Stroudsburg, PA: Dowden, Hutchinson and Ross, Inc.

Bernardi, T. 1946. "The Architect and the Housing Crisis: Custom Built Houses." *Progressive Architecture* (June): 52.

Blau, R.J. 1984. *The Architects and Firms: A Sociological Perspective in Architectural Practice*. Cambridge, MA: MIT Press.

Buchholz, R.A. 1993. *Principles of Environmental Management: The Greening of Business*. Paramus, NJ: Prentice-Hall.

Buwalda, J. 1980. "EMAD Expandable Minimum Ameriyah Dwelling." *Open House International* 2: 61–70.

Calthorpe, P. 1993. *The Next American Metropolis: Ecology, Community, and the*

American Dream. New York: Princeton Architectural Press.
Campbell, A. 1981. *The Sense of Well-Being in America: Recent Patterns and Trends*. New York: McGraw Hill.
Canada Mortgage and Housing Corporation (CMHC). 1997. *Housing Export Opportunities Series: Mexico*. Ottawa: CMHC.
– 1995. *Consumer Housing Preferences in the 1990s: An In-Depth Study of What Baby Boomers, Empty Nesters, and Generation X Want in Housing – Now and in the Future*. Ottawa: CMHC.
– 1994. *Building Prosperity: A Blueprint for Housing Export*. Ottawa: CMHC.
– 1991. *Strategic Plan: 1992–1996*. Ottawa: CMHC.
Canadian Manufactured Housing Institute (CMHI). 1998. *1997 Annual Report*. Ottawa: CMHI.
Clark, Jr, C.E. 1976. *The Early American Family Home*. New York: Oxford University Press.
Cooke, J.R. 1993. Speech delivered to the New Brunswick Mobile Home and Parks Association Annual General Meeting, 29 January.
Cooper Marcus, C., and W. Sarkissian. 1986. *Housing As If People Mattered: Site Design Guidelines for Medium-Density Family Housing*. Berkeley: University of California Press.
Cutler, V. 1947. *Personal and Family Values in the Choice of a Home*. Ithaca, NY: Cornell University Agricultural Experimental Station.

Dean, J., and S. Breines. 1946. *The Book of Houses*. New York: Crown.
Dear, M. 1992. "Understanding and Overcoming the NIMBY Syndrome." *Journal of the American Planning Association* 58, no. 3 (summer).

Eichler, N. 1982. *The Merchant Builders*. Cambridge, MA: MIT Press.

Ford, K.M., and T.H. Creighton. 1954. *Quality Budget Houses*. New York: Reinhold.
Fried, M., and P. Gleicher. 1961. "Some Sources of Residential Satisfaction in an Urban Slum." *Journal of the American Institute of Urban Planners* 27: 305–15.
Friedman, A. 1997. *La Casa a la Carta*. Montreal: McGill University School of Architecture Affordable Homes Program.
– 1992. *Design for Growth and Adaptability in Affordable Housing*. Montreal: McGill University Affordable Homes Program.
– 1991. "The Grow Home: Design Strategies for Low Energy Consumption – A Case Study." Proceedings of the 17th Annual Conference of the Solar Energy Society of Canada, 21–26 June 1991, Toronto, Ontario: 93–8.
– 1990. *Innovation and the North American Homebuilding Industry*. Montreal: McGill School of Architecture Affordable Homes

Program.
– 1987. "A Proposed Decision Making Model for Initiators of Flexibility in Multi-Unit Housing." Montreal: Unpublished PhD thesis, Université de Montréal.
– 1982. "Growth and Adaptability (G&A) in Housing." Unpublished M. Arch. thesis, McGill University.

Friedman, A., and V. Cammalleri. 1992a. *Evaluation of Affordable Housing Projects Based on the Grow Home Concept*. Montreal: McGill School of Architecture Affordable Homes Program.
– 1992b. "From Inception to Implementation: The Grow Home Experience." Paper presented at the Governor's Conference on Affordable Housing, Arizona State University, 16–18 January 1992.

Friedman, A., V. Cammalleri, J., Nicell, F., Dufaux, and J. Green. 1993. *Sustainable Residential Developments: Planning, Design and Construction Principles (Greening the Grow Home)*. Montreal: McGill School of Architecture Affordable Homes Program.

Friedman A., M. Horvat, and M. Rojano. 1997. *Adapting Quebec Construction Products to Latin American Markets*. Montreal: McGill University School of Architecture Affordable Homes Program.

Galster, G.C., and G.W. Hesser. 1981. "Residential Satisfaction: Compositional and Contextual Correlates." *Environment and Behavior* 13, no. 6: 735–58.

Gans, H.J. 1967. *The Levittowners: Ways of Life and Politics in a New Suburban Community*. New York: Pantheon.

Goethert, R., and Z. Shafie. 1979. "Housing for Low Income and Informal Sector." In *The Housing and Construction Industry in Egypt* (Cambridge, MA: MIT Interim Working Papers).

Gruber, K.J., and G.G. Shelton. 1987. "Assessment of Neighborhood Satisfaction by Residents of Three Housing Types." *Social Indicators Research* 19, no. 3: 303–15.

Gurstein, P. 1995. *Planning for Telework and Home-Based Employment: A Canadian Survey on Integrating Work into Residential Environments*. Ottawa: Canada Mortgage and Housing Corporation.

Ha, M., and M.J. Weber. 1991. "The Determinants of Environmental Residential Qualities and Satisfaction: Effects of Financing, Housing Programs and Housing Regulations." *Housing and Society* 18, no. 3: 65–76.

Habraken, N.J. 1976. *Variations: The Systematic Design of Supports*. Cambridge, MA: MIT Library of Architecture and Planning.
– 1972. *Supports: An Alternative to Mass Housing*. London: Architectural Press.

Harlap, E., ed. 1977. *Israel Builds 1977*. Israel: Ministry of Housing.

Hempel, D.J. 1976. "Consumer Satisfaction with the Home Buying Process: Conceptualization and Measurement." In H.K. Hunt, ed. *Conceptualization and Measurement of Consumer Satisfaction and Dissatisfaction*: Proceedings of a Conference Conducted by the Marketing Science Institute with Support of the National Science Foundation, 11–13 April 1976.

Herbert, G. 1984. *The Dream of the Factory-Made House: Walter Gropius and Konrad Wachsmann*. Cambridge, MA: MIT Press.

House and Home 1954. "Redesigned: Levitt Keeps Experimenting with the Expandable Attic" (February): 118–23.

Hudnut, J. 1942. "Education and Architecture." *Architectural Record* 92, no. 4 (October): 36–8.

Instituto Nacional de Estadistica, Geografica y Informatica. 1997. *INEGI: About Mexico*. www.inegi.gob.mx/homeing/acerca/acercademexico.html

Kantrowitz, M., and R. Nordhaus. 1980. "The Impact of Post-Occupance Evaluation Research: A Case Study." *Environment and Behavior* 12, no. 4: 508–19.

Kaynak, E., and L. Stevenson. 1981. *Home Buying Behaviour of Atlantic Canadians*. Halifax: Mount Saint Vincent University.

Kendall, S., and T. Chalmers. 1986. *Shell/Infill: A Technical Study of a New Strategy for 2x4 Housebuilding*. Cambridge, MA: MIT University Press.

King, A. 1990. "Narrow-Minded: Montreal Takes a Second Look at Impressive Housing Idea." *Montreal Gazette* 21 June.

Kjeldsen, M. 1988. *Industrialized Housing in Denmark*. Copenhagen: Danish Building Centre.

Kobrin, F.E. 1976. *The Fall in Household Size and the Rise of the Primary Individual in the United States Demography*. Vol. 13, no. 1: 127–38.

Krieger, A. 1991. "Since (and before) Seaside." In *Andres Duany and Elizabeth Plater-Zyberk: Towns and Town-Making Principles*, edited by A. Krieger and W. Lennertz (New York: Rizzoli).

Lansing, J.B., and R.W. Marans. 1969. "Evaluation of Neighborhood Quality." *AIP Journal*, May.

Leavitt, J. (1991) "Two Prototypical Designs for Single Parents: The Congregate House and the New American House." In *New Households: New Housing* edited by K.A. Franck and S. Ahrentzen (New York: Van Nostrand Reinhold).

Lynch, K. 1990. "The Openness of Open Space." In *City Sense and City Design: Writings and Projects of Kevin Lynch* edited by T. Banerjee and M. Southworth (Cambridge, MA: MIT Press).

MacDonald, D. 1996. *Democratic Architecture: Practical Solutions to Today's Housing Crisis*. New York: Whitney Library of Design.

Mandel, N., and A. Duffy. 1995. *Canadian Families: Diversity, Conflict and Change*.

Toronto: Harcourt Brace & Company Canada.
Miron, J.R. 1981. "The Two Person Household Formation and Housing Demand." Toronto: Canada Mortgage and Housing Corporation.
Mohr, M. 1989. "Small Packages, Big Ideas for High-Quality Low-Cost Houses." *Harrowsmith* (March): 66–7.
Montgomery, R. 1977. "High Density, Low Rise Housing and Changes in the American Housing Economy." In *The Form of Housing*, edited by S. Davis (New York: Van Nostrand Reinhold): 83–111.
Montreal Gazette. 1989. "McGill Centre Will Look at Ways of Making Housing Affordable." 16 February.
Morris, E.W., and M. Winter. 1978. *Housing, Family, and Society*. Iowa State University: Department of Family Environment.
– 1975. "A Theory of Family Housing Adjustment." *Journal of Marriage and the Family* 37, no. 1 (February): 79–88.
Mumford, L. 1961. *The City in History: Its Origins, Its Transformations, and Its Prospects*. New York: Harcourt, Brace & World.
Muthesius, S. 1982. *The English Terraced House*. New Haven: Yale University Press.

Nebel, B.J., and R.T. Wright. 1993. *Environmental Science: The Way the World Works*. 4th ed. Paramus, NJ: Prentice-Hall.
Nitaya, C. 1981. "Tung Song Hong: An Alternative Design Proposal." *Open House International* 6, no. 4: 30–40.

Onibokun, A.G. 1974. "Evaluating Consumers' Satisfaction with Housing: An Application of a Systems Approach." *Journal of the American Institute of Planners* (May).

Pastier, J. 1988. "A One-Architect Movement for Affordable Housing." *Architecture* (July): 64–71.
Pinero, G., and J. Slautsky. 1979. "Adaptable Housing in the Argentinian Context." *Open House International* 4; 62–76.

REIC Ltd. et al. 1991. *Making a Molehill out of a Mountain II: Implementing the Three R's in Residential Construction*. Ottawa: Canada Mortgage and Housing Corporation.
Reid, K. 1945. "Houses for the People." *Pencil Points (Progressive Architecture)* (September): 62.
Rent, G.S., and C.S. Rent. 1978. "Low-Income Housing: Factors Related to Residential Satisfaction." *Environment and Behavior* 10, no. 4: 459–88.
Rios, A.A. 1995. "Post-Occupancy Adaptation of Affordable Single-Family Housing in Montreal. Montreal." Unpublished M. Arch. thesis, McGill University.
Robbins, I. 1971. *Housing the Elderly*. Washington, DC: Technical Committee on

Housing.

Robinson, S. 1988. *Manufactured Housing: What It Is, Where It Is, How It Operates*. Barrington, IL: Ingleside Publishing.

Robinson, T. 1991. *Sustainable Housing for a Cold Climate*. Ottawa: Canada Mortgage and Housing Corporation.

Rogers, E. 1983. *Diffusion of Innovations*. New York: Free Press.

Rybczynski, W. 1989. "Architects Must Listen to the Melody." *New York Times* (Arts & Leisure, 24 September): 2–1.

Rybczynski, W., A. Friedman, and S. Ross. 1990. *The Grow Home*. Montreal: McGill School of Architecture Affordable Homes Program.

Schneider, F. 1994. *Floor Plan Atlas: Housing*. Germany: Birkhäuser Verlag.

Schoenauer, N. 1992. *History of Housing*. Montreal: McGill University.

Sears, Roebuck and Company. 1991. *Sears and Roebuck Catalog of Houses, 1926: An Unabridged Reprint*. Philadelphia: Athenaeum of Philadelphia and Dover Productions.

Senbel, M. 1995. "Working at Home and Sustainable Living: Architecture and Planning Implications." Unpublished M. Arch. thesis, McGill University.

Shankland Cox Partnership. 1979. *Site and Service Housing in Jamaica: The Camplands Project*. Kingston, Jamaica.

Sheltair Scientific Ltd. 1991. *Optimize: A Method of Estimating the Lifecycle Energy and Environmental Impact of a House*. Ottawa: Canada Mortgage and Housing Corporation.

Sherwood, R. 1978. *Modern Housing Prototypes*. Cambridge, MA: Harvard University Press.

Sichelman, L. 1987. "HUD's Plan for Affordable Housing Is Ready to Go into Mass Production." *Chicago Tribune* (6 September).

Stapleton, C.W. 1980. "Reformation of the Family Life-Cycle Concept: Implications for Residential Mobility." *Environment and Planning* A, no. 12: 1103–118.

Statistics Canada. 2000a. "Births and Birth Rate, Canada, the Provinces and Territories." www.statcan.ca/english/Pgdb/People/Population/demo04c.htm.

– 2000b. "Population, Occupied Private Dwellings, Private Households, Collective Dwellings, Population in Collective Dwellings, and Average Number of Persons per Private Household." www.statcan.ca/english/Pgdb/People/Families/fa mil66.htm.

– 1993. *Homeowner Repair and Renovation Expenditure in Canada*. Ottawa: Statistics Canada.

Sternthal, B. 1993. "Factors Influencing the Diffusion of Innovative Products in North American Home Building Firms." Unpublished M. Arch. thesis, McGill University.

Stevenson, K.C., and H.W. Jandl. 1986. *Houses by Mail: A Guide to Houses by Sears,*

Roebuck and Company. New York: Preservation Press.

Teasdale, P.L., and M.E. Wexler. 1993. *Family Dynamics, Residential Adjustments and Dwelling Adaptability*. Ottawa: Canada Mortgage and Housing Corporation.

Van der Ryn, S., and P. Calthorpe. 1986. *Sustainable Communities: A New Design Synthesis for Cities, Suburbs and Towns*. San Francisco: Sierra Club.

Varela Alonso, L. 1997. *Square Meter Construction*. Mexico: Bimsa Southam.

Wiedemann, S., and J.R. Andersen. 1982. "Residents' Perceptions of Satisfaction and Safety: A Basis for Change in Multifamily Housing." *Environment and Behavior* 14, no. 6: 695–724.

Wiedemann, S., A. Friedman, and W. Rybczynski. 1989. *Modular Prefabrication Versus Conventional Construction Methods As an Affordable Option in the Development of Single Family Housing*. Montreal: McGill School of Architecture Affordable Homes Program.

Wills, R.B. 1945. *Houses for Homemakers*. New York: Franklin Watts.

Wright, G. 1981. *Building the Dream: A Social History of Housing in America*. New York: Pantheon.

Zeisel, J., and P. Welch with C. Pilkington, M. Ertel, and P. Gill. 1981. *Housing Designed for Families*. Cambridge, MA: Joint Center for Urban Studies of MIT and Harvard University.

SOURCE CREDITS

Sections of some of the chapters in this book are based on previously published material. The references for these source materials are listed below.

Preface

Rybczynski, W., A. Friedman, and S. Ross. 1990. *The Grow Home*. Montreal: McGill School of Architecture Affordable Homes Program.

Chapter 1: Different Times, Different Homes

Friedman, A. 1999. "Thoughts on Homes and Consumption." *EnRoute*, September 32–8. Reprinted as "The High Price of Consumption," *Montreal Gazette*, 3 November 1999, B3.
Friedman, A., D. Krawitz, J. Fréchette, C. Bilimoria, and D. Raphael. 1996. *The Next Home*. Montreal: McGill School of Architecture Affordable Homes Program.

Chapter 2: Constructing Ideas

Friedman, A. 1995. "The Evolution of Design Characteristics during the Post-Second World War Housing Boom: The U.S. Experience." *Journal of Design History* 8, no. 2: 131–46.
Rybczynski, W., A. Friedman, and S. Ross. 1990. *The Grow Home*. Montreal: McGill School of Architecture Affordable Homes Program.
Friedman, A. 1994. "Narrow-Front Row Housing for Affordability and Flexibility." *Plan Canada* (September): 9–16.

Chapter 3: The Prototype

Rybczynski, W., A. Friedman, and S. Ross. 1990. *The Grow Home*. Montreal: McGill School of Architecture Affordable Homes Program.
Friedman, A. 1991. "Evaluating the Grow Home." Proceedings of the 1991 Annual Conference of the American Association of Housing Educators (AAHE), 64–72, University of New Hampshire, Durham, NH. (The author gratefully acknowledges the contribution of Donald Chan to the original research.)

Chapter 4: From Campus to Sites

Friedman, A., and V. Cammalleri. 1992. "From Inception to Implementation: The Grow Home Experience." Paper presented at the Governor's Conference on Affordable Housing, Arizona State University, 16–18 January 1992.

Chapter 5: Buying Their First Homes

Friedman, A., and V. Cammalleri. 1993. "Occupant Satisfaction with Narrow-Front Starter Homes in Montreal." *Housing and Society* 20, no. 1: 51–62.

Chapter 6: Making It Their Own

Rios, A., and A. Friedman. 1996. "Residential Modification of Narrow Front Affordable Grow Homes in Montreal, Canada." *Open House International* 21, no. 2: 4–17.

Chapter 7: The Grow Home of the Future

Friedman, A., and V. Cammalleri. 1997. "Cost Reduction through Prefabrication: A Design Approach." *Housing and Society* 24, no. 1: 1–4.

–1994. *Industrialization of Narrow-Front Rowhousing Using Wall Panel Systems: Industrialization of the Grow Home*. Montreal: McGill School of Architecture Affordable Homes Program.

Friedman, A. 2000. "How Will Home Building Change in the 21st Century?" *Hardware & Home Centre* (January/February): 31–2.

Chapter 8: La Casa a la Carta

Friedman, A. 1998. "La Casa a la Carta: North Helps House South." *Open House International* 23, no. 4: 4–12.

Chapter 9: Small Is Green

Friedman, A., and V. Cammalleri. 1994. "Reducing Energy, Resources and Construction Waste through Effective Residential Unit Design." *Building Research and Information* 22, no. 2: 103–8.

Chapter 10: Neighbourhoods with a Sense of Scale

Friedman, A. 1994. "Narrow-Front Row Housing for Affordability and Flexibility." *Plan Canada* (September): 9–16.

PHOTO CREDITS

Chapter 2
Figure 17: Avi Friedman

Chapter 3
Figure 6: Witold Rybczynski
Figure 7: Rick Kerrigan
Figure 8: Rick Kerrigan
Figure 9: Rick Kerrigan
Figure 10: Rick Kerrigan
Figure 11: Rick Kerrigan

Chapter 4
Figure 1: Avi Friedman
Figure 7: Avi Friedman

Chapter 5
Figure 1: Avi Friedman
Figure 2: *Montreal Gazette*
Figure 3: Avi Friedman
Figure 5: Avi Friedman

Chapter 6
Figure 5: Aurea A. Rios
Figure 7: Aurea A. Rios
Figure 8: Aurea A. Rios
Figure 10: Aurea A. Rios
Figure 12: Aurea A. Rios

Chapter 7
Figure 11: Jack Goldsmith

Chapter 8
Figure 1: Avi Friedman
Figure 2: Avi Friedman
Figure 9: Jack Goldsmith
Figure 10: Jack Goldsmith

Chapter 10
Figure 1: Avi Friedman
Figure 3: Avi Friedman
Figure 6: Avi Friedman
Figure 7: Avi Friedman

INDEX